INTOLERANT BODIES

JOHNS HOPKINS BIOGRAPHIES OF DISEASE
Charles E. Rosenberg, Series Editor

Randall M. Packard, *The Making of a Tropical Disease: A Short History of Malaria*

Steven J. Peitzman, *Dropsy, Dialysis, Transplant: A Short History of Failing Kidneys*

David Healy, *Mania: A Short History of Bipolar Disorder*

Susan D. Jones, *Death in a Small Package: A Short History of Anthrax*

Allan V. Horwitz, *Anxiety: A Short History*

Diane B. Paul and Jeffrey P. Brosco, *The PKU Paradox: A Short History of a Genetic Disease*

Gerald N. Grob, *Aging Bones: A Short History of Osteoporosis*

Christopher Hamlin, *More Than Hot: A Short History of Fever*

Warwick Anderson and Ian R. Mackay, *Intolerant Bodies: A Short History of Autoimmunity*

Intolerant Bodies

❋ ❋ ❋

A Short History of Autoimmunity

Warwick Anderson and
Ian R. Mackay

Johns Hopkins University Press
Baltimore

Printed in the United States of America on acid-free paper
2 4 6 8 9 7 5 3 1

Johns Hopkins University Press
2715 North Charles Street
Baltimore, Maryland 21218-4363
www.press.jhu.edu

Library of Congress Cataloging-in-Publication Data

Anderson, Warwick, 1958– author.
Intolerant bodies : a short history of autoimmunity /
Warwick Anderson and Ian R. Mackay.
p. ; cm. — (Johns Hopkins biographies of disease)
Includes bibliographical references and index.
ISBN 978-1-4214-1533-8 (pbk. : alk. paper) — ISBN 1-4214-1533-X
(pbk. : alk. paper) — ISBN 978-1-4214-1534-5 (electronic) —
ISBN 1-4214-1534-8 (electronic)
I. Mackay, Ian R., author. II. Title. III. Series:
Johns Hopkins biographies of disease.
[DNLM: 1. Autoimmune Diseases—history. WD 305]
RC600
616.97'8—dc23
2014006735

A catalog record for this book is available from the British Library.

Dedicated to
Barbara Gutmann Rosenkrantz
and
Ian Jeffreys Wood

CONTENTS

FOREWORD

Disease we have always had with us. Our ancestors suffered pains in their joints, debilitating coughs and exhausting diarrheas, sore throats and bloody urine, painful and sometimes mortal swellings. Ancient bones tell us that pathological processes are older than written records. And such written records tell us that there has never been a time when men and women have not elaborated ways of explaining the incidence and nature of such ills—often in connection with ideas about their prognosis and treatment. But older descriptions and terminology do not track easily onto twenty-first-century categories and understandings; such terms as *dropsy*, *continued fever*, *old age*, and *bloody flux* no longer populate our death certificates.

What, we are tempted to ask, were the "actual" ills from which our ancestors suffered and died? Today's scholars who are concerned with past morbidity and mortality cannot avoid starting with their own understandings, with a repertoire of specific disease entities. We have come, in the course of the past two centuries, to think of such entities as having a characteristic clinical picture, and of that picture as being the consequence of an underlying mechanism of some sort—whether it be the response to a microorganism, a malfunctioning kidney, or a genetic mutation. Where we have not yet agreed on such a mechanism, we assume that it will ultimately be revealed. And we have come to think of these entities apart from their manifestation in the bodies of particular individuals; they have become things that can be coded, arranged in orderly classifications, subjected to agreed-upon treatment protocols. Of course, each of these specific entities that we take for granted has a history and a geographical distribution related to that history. Thus the rationale underlying the Johns Hopkins University Press's biographies of disease. The very term *biography* implies a coherent identity and a narrative, a discernible movement through time.

And *biography* implies an ability to track that entity into the past through surviving sources—diaries, letters, medical treatises, clinical records. Classical descriptions in Greek and Latin reassure us that the regularly recurring fevers they document were in all probability what we would now recognize as malaria. Other ills are more elusive. Scholars have never agreed, for example, on the identity of the astonishingly mortal "sweating sickness" that scourged England in the late fifteenth and early sixteenth centuries. And some ailments visible to us would have been invisible to earlier generations—and therefore in the sources that survived them. If AIDS, to cite a conspicuous instance, had emerged in the ancient world or in the eighteenth century, it would have been invisible as an entity to contemporaries; men and women would have sickened, wasted, died, but in a bewildering variety of circumstances. AIDS, that is, could not have existed as a conceptual and linguistic entity until the biomedical tools of the twentieth century allowed it to be recognized as a pathophysiological phenomenon—with quite variable clinical manifestations.

In the present book Warwick Anderson and Ian Mackay have undertaken to write the "biography" of autoimmunity, a biological phenomenon literally inconceivable until the twentieth century, with its extraordinary developments in biomedical knowledge. By the last decades of that century, autoimmunity (in which the immune system fails in identifying the body's own tissues, fails to distinguish between self and non-self) was understood to be a mechanism underlying a variety of elusive chronic ailments, among them rheumatoid arthritis, systemic lupus erythematosus, multiple sclerosis, and type 1 diabetes. It is not that these ills did not afflict men and women before the second half of the last century. They were hardly invisible in the clinic or in the sickroom. Each of these ailments had attracted the attention of generations of practitioners; each had attained an identity based upon cumulative clinical experience. But as biomedical scientists in a variety of post–World War II laboratories began to elucidate the seemingly paradoxical phenomenon of autoimmunity, these diseases developed a more complex and multidimensional identity. Their individual biographies were altered and connected, based on their newly understood common origin in immune system dysfunction.

Disease is not simply a biological event, though it is often that, but, as Anderson and Mackay make clear, the brief history of autoimmunity provides an illuminating occasion for thinking about disease as multidimensional, and as time and place specific. In the laboratory, these related autoimmune pathologies are systems problems to be understood at the molecular and cellular level. For the clinician, on the other hand, they are a group of well-established, if frustratingly intractable, ailments to be diagnosed and treated. For sufferers and their families they are misfortunes to be experienced and endured. It is hard to avoid the conclusion that any particular autoimmune disease is in some sense an ever-changing composite of all these realities. These diseases exist not only at the molecular level in the laboratory and in journal articles and at specialty meetings but also as recurring and intractable realities to be experienced, diagnosed, and managed by patients and their families and by clinicians. There is no simple, one-dimensional way to understand those entities we call autoimmune diseases.

For an historian, the phenomenon of autoimmunity recalls an older way of thinking about the fundamental nature of disease, a way of thinking that was commonplace in 1800 but marginal by 2000. Today, many of us tend to think of disease as the outcome of a body's defensive reaction to a specific external pathogen or the working out of a genetic destiny predetermined—baked-in—at the moment of conception. Two centuries ago, however, the occurrence of disease was ordinarily understood by both physicians and laymen as an individual, physiological, response to an individual's unique bodily makeup and life course. Diet and exercise interacting with physical environment, life circumstances and constitutional endowment resulted in sickness or health. As the body moved through time it was continuously self-regulating, adjusting and readjusting and thus always at risk. In this traditional and nonspecific way of thinking about disease there was little role for infection and infectious agents. One might describe this way of thinking about the nature and origin of disease as holistic and physiological, as well as cumulative and biographical. The phenomenon of autoimmunity, with its implied juxtaposition of biological individuality and particular circumstance, reminds us of this older, integrative way of seeing the body in time and in terms of its cumulative interaction with itself and with its external—yet internalized—environment.

Autoimmunity is a resonant framework for thinking about identity in the labile world of postmodernity, and, as Anderson and Mackay so usefully indicate, the immune system has proved a seductive source of language and metaphor for a variety of philosophers and social theorists hoping to think with the body and about the body in society, to explore in contemporary terms ancient and enduring problems of the self and its relations.

Charles E. Rosenberg

INTOLERANT BODIES

INTRODUCTION

Thinking Autoimmunity

❋ ❋ ❋

Soon after waking one morning in winter 1981, a middle-aged writer, recently separated from his wife, fumbled as he tried to open a door in his Manhattan apartment. Later, when eating breakfast alone at a nearby restaurant, he experienced difficulty swallowing. His legs felt rubbery. He was unable to remove his sweater. "Something neurologically unpleasant was taking place inside me," Joseph Heller reflected. "Something I could not control and could not fathom."[1] As his limbs became weaker, the author of *Catch-22* called a physician friend, who quickly came over and diagnosed Guillain-Barré syndrome, a rapidly ascending paralysis of the peripheral nerves. Admitted to hospital, Heller soon could not sit up or support his head; his breathing was labored; the paralysis continued to spread. He spent weeks in the intensive care unit, but slowly he recovered, even though muscle weakness would linger one more year. At the time Heller received his diagnosis, Guillain-Barré syndrome had been recognized for decades as an autoimmune disease.[2] It is one of more than eighty separate diseases caused when the immune system attacks the body's own tissue—when an individual's normal defensive processes go askew and turn pathological. "Some small biological process in my body, for reasons still unknown to anyone," Heller later wrote, "began manufacturing . . . antibodies with a pathological affinity for infiltrating and degenerating the tissues surrounding the fibers in my peripheral nervous system."[3] The writer's body had gone immunologically haywire.

Although figures are uncertain, it appears that some 5 to 10 percent of any population, most of them women, can expect to suffer from an autoimmune disease in a lifetime.[4] Guillain-Barré syndrome is rare, affecting perhaps one in 100,000 people each year worldwide, but other autoimmune conditions, such as multiple sclerosis, systemic lupus erythematosus, rheumatoid arthritis, and type 1 diabetes mellitus are more common.[5] Most of these ailments, unlike typical Guillain-Barré, cause persistent suffering, following a drawn-out, often lifelong, pattern of remission and recurrence, of control and exacerbation. Together they represent a major disease burden—and a rapidly growing global health problem, as the number of cases is increasing, for reasons still unclear. Yet, autoimmunity, as a mechanism of disease causation, was almost unthinkable until the 1950s. Indeed, there is no record of the use of *autoimmune* as an adjective before 1951, and the term *autoimmunity* was coined only in 1957.[6] Although scientists had studied the body's immunity to foreign agents since the late nineteenth century, most experts were unable to contemplate its defensive contrivances turning rogue, going on the offense against their own body. A few investigators did wonder in the early twentieth century about the dangers of immunological overreactivity or hypersensitivity, about allergy and anaphylaxis; but they hesitated to propose autoimmunity as a valid, or at least consistent, mode of disease causation. It seemed too dysfunctional to be plausible. After World War II, autoimmunity would gradually gain adherents, though it remained a stubbornly marginal, and even far-fetched, notion in modern medicine.[7]

The paradox at the heart of autoimmunity helps to explain the durable resistance to the concept. While most people will fastidiously guard against microbial invaders, understand the threat of toxins, watch apprehensively for cancer, foresee heart disease and stroke, and readily explain genetic disease, they find the idea of betrayal by the immune system curiously disturbing and repellant. That the immune system, so much a part of us, so necessary to survival, can go amiss and cause disease is counterintuitive. The body's failure to recognize itself, its capacity to treat itself as foreign, seems both sinister and bizarre. Thus, even as immunologists and physicians belatedly and sometimes reluctantly took up the concept of autoimmunity, many patients—Heller is an exception—continued to find the idea perplexing. As a pathological process, autoimmunity is still emerging, still to gain broad cultural acceptance. Only recently has autoimmunity begun to find its voice in public.

Some may say it is too soon to write the history of autoimmunity, yet autoimmunity has a long intellectual genealogy and the diseases it explains are old and well-defined afflictions. In a sense, autoimmunity is the current manifestation of the traditional medical concern for idiosyncrasy or individual variation in the causation and expression of disease. Its very definition includes sediments of earlier interest in disease as a biographical process, interest in how normal function might shade into pathology, in how illness derives from, and articulates, human individuality.

Until late in the nineteenth century, most medical doctors—and most of their patients—regarded disease as a disturbance of an individual's constitutional equilibrium, ensuing from imbalance in consumption and excretion, a mismatch in the interaction of body and environment. Hereditary disposition, or diathesis, might render an individual prone to one disorder or another, but whether this tendency actually was excited into disease depended on circumstances and habits. The body was perpetually in a state of flux, sensitive to changes in diet, behavior, clothing, environment, climate, and so on. Thus disease, often expressed as fever, indicated some physiological disruption in the individual, and its cure required physiological regulation, through either stimulatory or depletive measures. But the development of germ theories in the nineteenth century dissolved much of this constitutional model. Increasingly, disease appeared to derive from external entities, from microbes cultivable in the new laboratories. Less the disturbance of a physiological process, disease came to assume a standard modular form, that is, to become "ontological"; concrete nemesis thus entered the world of etiology, or disease causation.[8] Accordingly, medicine became less personalized, less dependent on biographical inquiry and individual monitoring, and more committed to tracking down, suppressing, and expelling specific microbes responsible for specific diseases. So goes the conventional story. In fact, clinical entanglements continued in the twentieth century to draw attention to idiosyncrasy and variation in disease causation and expression, especially in the domains of mental illness, cancer, allergy, and autoimmunity.[9] While critics of twentieth-century medicine like to depict it as an arid wasteland of reductionism, standardization, and routinism, there were hidden seams running through it, segments containing rich, variegated deposits of thought about individual reactivity and sensitivity. The story of autoimmunity, then, is nothing less than the history of concepts of biological individuality within twentieth-century medicine, an alternative history of biomedicine.

❋ ❋ ❋

In a fashion, this book tells the biography of autoimmune disease.[10] It describes the antecedents, origins, growth, and ups and downs of the relatively brief career of autoimmune disease. But there are many different autoimmune diseases, and here we focus on only four—multiple sclerosis, systemic lupus erythematosus, rheumatoid arthritis, and type 1 diabetes mellitus—though many others come into the field of vision. Perhaps more aptly and evocatively, our history of autoimmunity is not so much the collective biography of this family of diseases as the biography of a concept of pathogenesis, even the biography of a biographical framing of disease. Of course, such life stories tend to assume a certain end point. Even as they reveal a contingent and intricate conceptual itinerary, an erratic trajectory, biographies of this sort ultimately offer confirmation of our contemporary intellectual verities. In a sense, we are tracing a genealogy of the present—the present as experienced by people with autoimmune disease, those who care for them, and others who seek a scientific explanation for what ails them.[11] We want to see how autoimmunity became thinkable.

We have tried to write a conceptual history—or if you like, an epistemological history—of autoimmunity, at the same time as we seek to root the concepts in the experimental practices and clinical activities that nourish them.[12] In this way, we consider how concepts of autoimmunity developed and became operational in different settings. Some sites favored the growth of autoimmune knowledge more than others. Notions of individual reactivity and sensitivity flourished best where the prevailing thought styles and sensibilities nurtured such ideas. Thus, the concept of autoimmunity cannot be removed from the circumstances in which it initially developed—whether in Paris, Vienna, New York, or Melbourne. Ideas of autoimmune pathology are also bound up with particular material and technical changes in scientific investigation and clinical pursuits. Without the invention of new laboratory techniques, the rise of institutions connecting laboratories with hospital wards, and the expansion of medical research funding after World War II, autoimmunity would have remained inconceivable. The widespread dissemination and stabilization of concepts of autoimmunity was contingent on the spread around the world in the late twentieth century of facilities for clinical research. Moreover, we cannot ignore the complicated and productive entanglement of ideas of autoimmunity with philosophical and sociological concepts of identity and self—and the later influence of cybernetics, communications theory, and operations research. It may be said that some aspects of the autoimmune schema embody a style of Cold War thinking—especially its emphasis on surveillance, recognition, control, conformity, and regulation. These and other factors, as we shall see, contributed to making the discovery and management of autoimmunity into global projects in the late twentieth century.

Additionally, this book is an attempt to describe the interpretation of disease—whether by scientists, physicians, philosophers, or patients—as intellectual practice. We thus treat as philosophical inquiry what others may dismiss as laboratory tinkering; we take seriously the reasoning in clinical activities; and we listen closely and respectfully to sufferers' explanations of their conditions. While we concentrate here on how certain puzzling, and usually chronic, diseases became framed as autoimmune disorders, we recognize that concomitant reflection on self and identity, and the normal and the pathological, inevitably turns into a *forme fruste* of philosophy of biology or implicit social theory. In speculating on autoimmunity, scientists and clinicians are reshaping conceptions of the normal and the pathological. In dealing with debilitating illness, sufferers like Heller are thinking about the constitution of self, doing biographical work, and wondering about the significance of their infirmity—even if they frequently feel that the concept of autoimmunity fails to capture fully their experiences. In contrast, autoimmunity has growing appeal to philosophers and social theorists as a guiding metaphor in understanding the perils of life and identity in the twenty-first century. In our own times, leading scholars in the humanities have taken the immunological turn. Accordingly, we expand what counts as a concept, and who counts as a thinker, in the "conceptual history of disease" in order to include all of these intellectual practices and practitioners.[13]

As clinicians, we understand only too well that autoimmunity is much more than an intellectual practice or thought style.[14] Its most potent meanings are inherent in the disease process and the experience of suffering, which both surpass our current understanding. Certainly they escape the words we try to find to contain them. "My blood plasma was filled with an antibody that destroyed peripheral nerve cells," writes American poet Sarah Manguso.[15] A Harvard undergraduate, she awoke one morning feeling numb, struggling to breathe. She stumbled and fell when she tried to get out of bed. A very rare chronic form of Guillain-Barré syndrome was taking over her body. Learning to live with her disease, she realized that "all autoimmune diseases invoke the metaphor of suicide. The body destroys itself from the inside." After several years, this autoimmune affliction went into remission. "The only thing I'd done in life was recovering from a disease," she writes. "My self-image had been highly susceptible to that event. It constituted most of my identity." As a budding writer she felt an abiding frustration. "How can I stop thinking about the disease long enough to write about anything else? How can I stop thinking about everything else long enough that I can write about the disease?"[16]

CHAPTER ONE

Physiology with Obstacles

⁂

Fever was the cardinal disorder of the nineteenth century—one might even say it was pathognomonic, or characteristic, of the times. People commonly suffered and died from intermittent or continued fevers; some succumbed to brain fever or puerperal fever; others could be admitted to fever hospitals with scarlet, typhoid, or yellow fevers; doctors might prescribe them fever bark. Fevers were described as putrid, malignant, or nervous. Things happened at fever pitch or as though in a fever dream. Some places were fever nests. It was necessary to guard against colonial fever and ship fever, to ward off camp fever and jail fever, to endure seasoning fever, and to sniff through hay fever and spring fever. Gold fever regularly excited millions. Cabin fever depressed others. In the 1840s, Edgar Allan Poe wrote about "the fever called 'Living' that burned in my brain." For Charles Dickens it was a "fevered world."[1] Of course, fever has always been with us, but in the nineteenth century it excited more medical, literary, and social interest than ever before. Fever was the key manifestation of a morbid condition in any system, whether personal or political; preternatural heat and irritability implied excessive change and destruction of the tissues of individuals and societies.[2] It happened when normal functions and regulatory mechanisms went awry, when the system became overexcited and began to damage itself. Notwithstanding clinical differences, fever in the nineteenth century therefore was the conceptual equivalent of autoimmunity, the signal pathology of the late twentieth century.

A destructive reaction of the immune system against the body's own tissues, autoimmunity became imaginable only in the twentieth century. Even after immunological processes were postulated, it still was hard to believe they might actually cause disease as well as resist external pathogens. But, despite apparent novelty, autoimmunity possesses many conceptual affinities with older explanations of pathogenesis, theories that identified dynamic internal derangements of function or physiology as causes of disease. In our time, autoimmunity has acted as a cognitive substitute for nineteenth-century notions of essential fever, offering yet another way of talking about abnormalities or exaggerations of bodily function. Accordingly, the conceptual history of autoimmunity—not primarily a febrile condition—nevertheless has a feverish beginning.

❋ ❋ ❋

Fever, as novelists observed, changed people and shifted boundaries. For Charles Dickens, bouts with fever occasioned his characters' personal crises and made available to them an opportunity for reassessment of self and other. In *Great Expectations*, Pip suffered from brain fever after the death of Magwitch:

> That I had a fever and was avoided, that I suffered greatly, that I often lost my reason, that the time seemed interminable, that I confounded impossible existences with my own identity; that I was a brick in the house-wall, and yet entreating to be released from the giddy place where the builders had set me; that I was a steel beam of a vast engine, clashing and whirling over a gulf, and yet I implored in my own person to have the engine stopped, and my part in it hammered off; that I passed through these phases of disease, I know of my own remembrance, and did in some sort know at the time.[3]

Pip experienced fever as discomposing his sense of himself and unsettling personal boundaries. And yet, nursed by Joe Gargery, he would emerge from the sick room more mature and self-confident, a transformed man.

Poor Esther Summerson, afflicted in *Bleak House* with smallpox, also felt her social self turn strangely contingent and unreliable in her feverish condition. She imagined herself in a "flaming necklace strung together somewhere in great black space." Irritable and unable to rest, "divisions of time became confused with one another. . . . At once a child, an elder girl, and the little woman I'd been so happy as, I was not only oppressed by cares and difficulties adapted to each station, but by the great perplexity of endlessly trying to reconcile them."[4] Fever confounds past and present, blurs social distinctions, and disturbs the whole body. Mr. Peggoty in *David Copperfield* tells us that Little Emily "took bad with fever." She felt "no today or yesterday, nor yet tomorrow; but everything in her life as ever had been, or ever could be, and everything was as never had been, and as never could be, was a crowding on her all at once, and nothing clear nor welcome."[5] Fundamentally, these are adaptation narratives in which people lose themselves in fever and then, if they are lucky, refashion a workable persona. As critic Miriam Bailin notes, "the Victorian sickroom scene, at its most typical, serves as a kind of forcing ground of the self—a conventional rite of passage issuing in personal, moral, or social recuperation."[6] Although high fever is experienced quite differently from chronic autoimmune conditions, they share this influence on identity and sense of self.[7]

❋ ❋ ❋

The concepts of fever in the nineteenth century are as intriguing as its multiple experiential registers—and especially pertinent to later ideas about autoimmunity. Fever implied that the whole body and its constitution were in turmoil. It seemed a manifestation of systemic disturbance and therefore prompted thoughts of significant alteration in identity, even of the reconstitution and reformation of the individual. The attribution of inciting cause, or etiology, remained perplexing, with some medical experts favoring the general mismatch of body and circumstances and others a specific inflammation with secondary systemic effects. Fever might represent the disease itself or be a symptom of efforts to restore balance in the body. It could develop out of a nervous insult or a circulatory commotion or an intestinal disorder.[8] Predictably, it was brain fever—nervous debility or prostration—that most fascinated nineteenth-century novelists and the reading public. Thus, when Heathcliff returned to Wuthering Heights, Catherine Linton showed signs of "brain fever." After Emma Bovary learned that her lover Rodolphe had abandoned her, she languished with "une fièvre cérébrale." Ivan in *The Brothers Karamazov* developed brain fever from his intense mental preoccupation.[9] When grief and anxiety and fatigue disordered delicate constitutions, fever was the frequent result. As a systemic complaint, whatever its causes, fever required holistic interventions. Since the bodily constitution usually appeared overexcited, this meant depletive measures, such as blood letting. Rest and supportive care also would help to subdue the fire within.

For many physicians, it was a matter of sensibility and irritability. The eighteenth-century Scottish physician William Cullen had argued persuasively that nerves were dominant in determining health and disease. This meant, in effect, that most fever, or pyrexia as he called it, was fundamentally neurotic, that is, related to the nervous system. Admittedly, some fevers could be associated with obvious inflammation at various locations, but in general they seemed akin to "brain fever," and thus interrelated. Fever was nature's signal of distress, evidence of excessive nervous irritability.[10] One of Cullen's students, John Brown, elaborated on his mentor's theories, claiming that life was a state maintained by stimulation, whether external, internal, or emotional. According to Brown, all disease therefore was the manifestation of either excess or deficiency in excitation; disease reflected either *sthenia* (overexcitability) or *asthenia* (exhaustion).[11] The London physician George Fordyce read Cullen with more skepticism, tending to doubt the "metaphysical reasoning" of dour Scots and flamboyant Frenchmen, and he argued that fever must be studied "in the bodies of men afflicted with this distemper." Fordyce believed that fever could derive from "infectious matter," exposure to cold, bad food, moist clothes, and sudden passion—so it sometimes helped to evacuate noxious materials from the body after their concoction.[12] Henry Clutterbuck was more confident in assigning all primary fevers to brain inflammation—though he too recommended "prompt and judicious employment of blood-letting."[13]

Later in the nineteenth century, it became more common for physicians to attribute fever to some essential disturbance or concoction of the blood, with the heat rising when the blood became morbific as it moved around the body. Thus, John Thomson, professor of surgery at Edinburgh, warned of the dangers of increased blood flow and noted sadly the "buffy coat" or "inflammatory crust" of the blood of fever cases.[14] Others described the "excessive richness" of the blood in fever.[15] Internal factors, against the background of predisposition and temperament, must have caused the disturbance, not any external agent. Whether something in blood began to ferment and become corrupted or whether simple friction produced the symptoms remained subject to dispute. In any case, it clearly was necessary to correct this disorder of vital fluids—this physiological derangement—through bleeding, sweating, and purging the patient. Since fever indicated some functional problem, the body needed to be rebalanced through enhancing its secretions and excretions.

Despite continued enthusiasm, especially in Britain, for physiological explanations of preternatural heat, French physicians were attempting to locate more securely the "lesion" of each disease, its exact site in the tissues of the body. Some luminaries of the Paris hospitals, like Xavier Bichat, insisted that specific tissue pathology could be identified for every ailment. For them, fever would mostly connote some underlying tissue defect or inflammation of organic matter (making it red, hot, swollen, and painful)—in this sense all fever was authentically "secondary" and only deceptively "primary" or "essential." Others combined their obsession with the lesion with an abiding interest in functional aspects of pathology. In the early nineteenth century, the liberal physician

François J. V. Broussais aggressively promoted his theory that all fever derived from inflammation and irritation of the gastrointestinal tract, a "lesion" or incitant that caused general disturbance of bodily function. He condemned the "ontologies" of his predecessors—their symptom-based clinical classifications—and favored instead a broader physiological explanation of disease, with an emphasis on the irritability or sensibility of tissues. For him there could be no specific diseases, just the systemic consequences of inflammation, mostly located in the gut.[16] Pathological conditions were simply exaggerations of the normal state, and therefore physiological disorders.[17] In thus elaborating on Brown's earlier speculations, Broussais gained an international following. A physician at St. Mary's Hospital, London, C. Handfield Jones, made a pointed distinction between asthenic fevers, which resulted from nervous exhaustion, and sthenic or inflammatory fevers, in which "the blood, being hotter than natural, stimulates the heart unduly."[18] D. J. Corrigan, the physician to Queen Victoria in Ireland, declared that fevers were "lesions of function."[19]

But what influenced the character and course of fever? How did generic physiological disorders acquire distinctive form and duration? Evidently, bodily constitution and temperament, shaped by heredity, would be important factors in any functional explanation of fever. Physicians recognized the individuality or idiosyncrasy of bodily systems, and their propensity to change over time, as shaping disease patterns. Diagnosis therefore included a biographical component. Brown had merely tried to simplify the influences on disease expression when he postulated sthenic and asthenic diatheses or predispositions. During the nineteenth century, diathesis was further disaggregrated and specified, so that a patient might display inflammatory or rheumatic or consumptive or any of a hundred other morbid temperaments or physiological inclinations.[20] But such concern with terrain and temperament would decline late in the century after the discovery of pathogenic germs, which offered a more compelling ontological explanation of many diseases—where diseases are separate entities that breed true—

eventually displacing the interest in personal diathesis onto fields like allergy and, later still, medical genetics and clinical immunology.

Torn between ontological and physiological theories of disease, and between secondary and primary origins for fever, most physicians resorted to skepticism, observation at the bedside, and pragmatic intervention. Ordinary doctors for most of the nineteenth century repeatedly claimed that any attempt to assign a cause to fever, or determine its relation to inflammation, was premature. "Amidst the varied difficulties which assail the young practitioner at the outset of his career," wrote an assistant surgeon in 1849, "I know of none more perplexing—and more disheartening—than those he meets with, at every turn, in his dealings with the inflammations." He worried that so few of his colleagues escaped the "infection of Broussaism and Clutterbuckism."[21] A medical lecturer at Guy's Hospital, London, complained that the "alien pellucidity" of medical theories was estranging them "from one's practical middle-class mind." Indeed, "in dealing with living nature our language tends to put us mentally athwart the facts of the case."[22] Intense observation on the wards was the answer.

Analysis of the blood of fever cases, shorn of any etiological speculation, might also offer some empirical insight into the character, if not the underlying meaning, of fever. In 1860, Alexander Tweedie, the physician to the London Fever Hospital, saw all fever "as originating in certain changes in the blood"; therefore, it was necessary to find a "means of counter-acting or neutralizing the effects of the fever poison." "The duty of the practitioner is to endeavour to guide the disease," he wrote, "and to prevent as much as possible injury to the organs essential to life."[23] Benjamin Travers, one of Queen Victoria's surgeons, also believed that a "change in the properties and distribution of the blood" was the mechanism of action of fever and inflammation.[24] Evidently, the chemical researches of Justus von Liebig had impressed some of these medical experts. Liebig argued that organic molecules are unstable, tending toward decomposition, which manifests as fermentation, putrefaction, and decay. As a complex fluid, blood would be especially susceptible to chemical decomposition, leading to a process of fermentation, expressed as fever.[25] Soon discarded for microbiological theories, physico-chemical explanations of one sort or another briefly proved popular. Donald Maccalister in his Gulstonian Lectures on the "intimate nature" of fever marveled at the normal constancy of body temperature, suggesting, "It rests on a perpetual balance of opposing tendencies which is as mysterious and as beautiful as anything in our frame." But then the "excessive heat and the excessive waste products of fever are evidence that the fabric itself is being wastefully consumed."[26] Another physician, writing on the eve of the acceptance of germ theories, wrote, "We are of necessity compelled to look upon fever as a disorder of the protoplasm."[27]

Fever often excited intense philosophical speculation during the nineteenth century—just as autoimmunity would in the twenty-first century. Georg W. F. Hegel, for example, thought fever represented the "pure life of the diseased organism." The idealist philosopher imagined illness as the mismatch of living ego and bodily reality, the "disproportion of its being and its self."[28] He described fever beginning with

> shivering, heaviness of the head, headache, twinges in the spine, twitching of the skin, and shuddering. In this activity of the nervous system the muscles are left free and consequently their own irritability functions as an uncontrolled trembling and powerlessness. . . . The organism dissolves all its parts within itself in the simplicity of the nerve, and feels itself withdrawing into the simple substance [of its being].

Soon this activity is transformed into "*heat, negativity*, where the blood is now the dominating factor."[29] Eventually, profuse sweating might restore healthy identity. "Sweat is the critical secretion; in it, the organism attains to an excretion of itself, through which it eliminates its abnormality and rids itself of its morbid activity." For Hegel, fever was thus the effort toward restitution and redemption. "Even if fever is, on the one hand, a morbid state and a disease, yet on the other hand, it is the way in which the organism cures itself."[30]

What, then, do we mean when we claim that theories of fever anticipated the idea of autoimmunity? The historical distinction between ontological and physiological explanations of disease is perhaps most compelling. These two analytic modes also echo differences between solid and humoral or static and dynamic or local and systemic frameworks. Historian of medicine Erwin H. Ackerknecht distinguished between the ontological view, "which sees the disease," and the individual approach, "which sees the patient."[31] According to German pathologist Rudolf Virchow, writing in the middle of the nineteenth century, diseases "represent only the course of physiological phenomena under altered conditions"; pathology is "nothing but physiology with obstacles."[32] For much of the nineteenth century, such functional explanations, often invoking bodily constitution, individuality, and diathesis, prevailed in accounts of fever and other human pathology. Disease—*dis-ease*—implied an exaggeration or alteration of the normal state, not an opposed condition. Thus, fever seemed an internal disorder or dysfunction, a physiological or regulatory mechanism gone awry—just as autoimmune disease would later appear. Only toward the end of the nineteenth century did fever and other pathologies commonly assume distinct ontological forms, generally attributed to microbiological agency; that is, clinical features would develop into discrete packages organized around specific invading germs. When they saw a patient, physicians then came to look for the disease, not the person.

As the concept of autoimmunity emerged in the middle of the twentieth century, it therefore represented to a degree the return to older physiological or biographical concepts of disease causation—to what might be imagined as the prelapsarian era of pathological vision, before the bacteriological fall, when evil, in the form of germs, entered the world of etiological theory.[33] Although immunology would arise from bacteriological investigation, autoimmune disease actually is predicated on the idea of disease as a biographical process, a personalized pathology. Diagnosis has to address the physiological function of the person and its potential dysfunction as it goes about the normal processes of preserving or restoring health.[34] In this sense, autoimmunity has the ambivalent status possessed by fever in the nineteenth century, as it represents—in fact, quite vividly—a destructive aspect of the body's regulatory mechanism. It raises the question of how to determine when imbalance or deviation actually becomes disease. This makes autoimmunity an exceptionally revealing clinical category, one more akin to complex, progressive hereditary conditions—modern diatheses—than simple infectious disease.

The analogy of autoimmunity with nineteenth-century concepts of fever can be pushed further. As our survey of fever theory indicates, the emphasis on systemic derangement, on organismic dysfunction, inevitably turns attention to issues of bodily integrity and the sense of self. Constitutional maladies of this sort tend to reveal individual proclivities and temperament and to have potential for refashioning bodily boundaries and social position. Perceived as disorders of normal function, fever and autoimmunity both would prompt intimate review of identity. Moreover, together they help to illuminate professional tensions between those favoring such speculative interpretations and their skeptical colleagues who demand more intense and sober observation of individual cases. Although manifested quite differently, both fever and autoimmunity thus can elicit philosophical conjecture and cautious empiricism, with each mode of practice appealing to remarkably stable competing professional self-images. Additionally, scrutiny of the concepts of fever and autoimmunity shows us the shifting fortunes of integrative aspirations and reductionism in medical thought, particularly the repeated tendency to substitute biochemical mechanisms for complex biological theories. Virchow in 1855 declared, "It is high time that we give up the scientific prudery of regarding living processes as nothing more than the mechanical resultant molecular forces inherent in the constitutive particles." As he later put it, "life is different from processes in the rest of the world, and cannot simply be reduced to physical or chemical forces."[35] In the middle of the twentieth century, the more biologically minded physicians and scientists again would be heard ruing the irresistible incursions of a different generation of biochemists with molecular toolkits.

❋ ❋ ❋

With the arrival of germ theories, or bacteriology, in the late nineteenth century, our febrile analogy cools off. Most fevers soon acquired their causative microbe, an external pathogenic agent that seemed to give the disease a robust and standard form more or less independent of the individual sufferer. Fever became less a functional disturbance of the sick person than a standard symptom complex indicating specific infection. As historian Henry E. Sigerist suggested in 1932, "man falls into the background compared to disease. Systematizing is no longer concerned with man but with disease."[36] The roots of this microbiological revolution were deep: speculations about the existence of a contagium, usually presumed inanimate, had circulated for centuries; the microscope had long ago revealed tiny creatures; fungus had seemed to blame for some skin disease even in the 1830s; and about the same time, living yeast was discovered to be the cause of alcoholic fermentation. All the same, it was hard at first to place these observations in the thought style of early nineteenth century medicine. Conceptual resistance eventually tumbled, however. Investigating fermentation, French chemist Louis Pasteur determined in the 1850s that a number of microbiological agents could cause wine and beer to spoil. These microbes appeared to breed true and cause predictable results. Their significance for human disease was becoming clear. In Scotland, the surgeon Joseph Lister decided that germs might cause wound infection, and sure enough, disinfecting dressings improved clinical outcomes. In the 1870s, German physician Robert Koch confirmed that a rod-shaped germ, the bacillus, was responsible for anthrax; and within a decade he identified the microbiological causes of those great scourges of humanity, tuberculosis and cholera. At the end of the nineteenth century, microbe hunters were revealing one after another the minute life forms responsible for dozens of major diseases.[37]

Yet, some chronic and systemic diseases seemed to defy bacteriological inquiry. Commonly, there was no shortage of candidate germs, but none of them was securely attached to these enigmatic diseases for long. Diathesis and nonspecific inflammation often retained a rather dated plausibility in explaining many chronic conditions. Examples are manifold, but it is particularly helpful to focus here on four illustrative ailments: rheumatoid arthritis, systemic lupus erythematosus, childhood (type 1) diabetes, and multiple sclerosis. The character of each disease is distinct and their conceptual histories are diverse, but all of these conditions prompted recurrent thoughts of predisposition and systemic disorder, and later, as we shall see, they lent themselves to a surprising immunological explanation.

Until the middle of the nineteenth century, gout and acute and chronic rheumatism clustered together as manifestations of "arthritism" or rheumatic diathesis; as afflictions more generally of asthenic or lowered constitutions. Starting in the 1850s, Alfred Baring Garrod tried to differentiate from this clinical amalgam the condition he called rheumatoid arthritis, a separate disease characterized by chronic inflammation of multiple joints, mild fever, general impairment of vigor, distortion of fingers, and deformity of knuckles. While this ailment was readily distinguished from gout, which demonstrated sudden and more contained attacks and elevated uric acid levels, its relation to acute rheumatic fever, characterized by high temperatures and migratory polyarthritis, seemed less than clear. Surveying the field from King's College Hospital, London, Garrod believed they comprised different diseases; but others, like Jean-Martin Charcot, based at the vast Salpêtrière hospital in Paris, regarded them as arbitrary points along a rheumatic continuum, as slow or acute expressions of the same propensity.[38] Whether rheumatoid arthritis was an emerging disease, excited by the Industrial Revolution, or just a novel category of disease derived from ever more sophisticated medical discourse also elicited considerable debate.

Whatever its cause, rheumatoid arthritis was diagnosed more frequently as the century rushed to its close. In the 1890s, Alfred Garrod described some of the cases he was observing. He saw an eighteen-year-old girl, tired, weak, with recent pain and swelling in her ankles, hands, and elbows. She could scarcely move her left elbow. But, following a rich diet and copious stimulants, she eventually showed some improvement. Another case, a married woman thirty-three years old, presented after months of pain, swelling, and stiffness in her fingers and wrists, worse in the mornings. She, too, after a tonic treatment seemed to suffer less. A man aged fifty-five turned up when his hands gradually became stiff and painful. As his other joints eventually stiffened, he found he could hardly move until late in the afternoon. "In such a case," Garrod wrote, "all lowering treatment is sure to be followed by an aggravation of the disease, and the only chance of arresting it is the steady perseverance in a proper tonic medicinal treatment and diet, and, in fact, everything should be done to sustain the general health at the highest possible pitch."[39] Thus did the older physician's casebook fill with instructive accounts of the rebalancing of rheumatic constitutions or diatheses.

In keeping with the new fashion for bacteriological investigation, Garrod's son Archibald briefly wondered in 1890 if some malignant bacterium might be responsible for rheumatoid arthritis, although generally he favored an internal neurological origin.[40] Others became more adamantly microbial, even when no organism could be isolated reliably from affected joints. Since the 1880s, the gonococcus, the agent of much venereal disease, was known to cause septic arthritis; and in the early 1900s, the connection between a bacterial infection of the throat and the development of rheumatic fever was established.[41] Surely then, some bug would be found for rheumatoid arthritis. As a sort of compromise, William Hunter in 1901 proposed his theory of "focal sepsis," whereby an infection elsewhere in the body might liberate toxins that targeted joints. In Chicago, Frank Billings successfully promoted focal infection, convincing most rheumatologists—a new type of specialist in medicine—that streptococcal infiltration of the tonsils was especially pathogenic.[42] By the 1930s, however, bacteriological skepticism would again prevail.

Such etiological speculation meant little to sufferers like Pierre-Auguste Renoir. From the 1890s, aged in his fifties, the painter became increasingly handicapped. Rheumatoid arthritis deformed his hands to the point where his wife or son had to fix the brush under his twisted fingers. Renoir's right shoulder stiffened and then froze. In his sixties he required one walking stick, then two; and by age seventy, he was confined to a wheelchair. Rheumatic cachexia caused him to shed weight. Antipyrine—a salicylate anti-inflammatory medication—and frequent purges failed to arrest his decline. He tried physical exercise, even juggling in the earlier phases, and visited spas, but nothing gave relief. Through all this, he never could confidently assign responsibility for his torment.[43]

Lupus is another disease that gained visibility and definition in the mid-nineteenth century, while remaining a pathogenic puzzle. Young women increasingly presented to physicians with inflamed cheeks and face, with red patches distributed over the skin in the shape of a butterfly's wings, and sometimes a low fever. Then called seborrhea congestiva or inflammatio folliculorum or érythème centrifuge, the condition initially resembled a slow tuberculosis-like affection, and was thought therefore perhaps to be an expression of a scrofulous or consumptive diathesis—or alternatively, one of the protean manifestations of syphilis, or even a form of cancer. At first, this condition caused considerable diagnostic confusion. Around 1850 the French dermatologist Pierre L. A. Cazenave named the disease "lupus érythèmateux," attributing it to a lymphatic constitution—not to consumption (or tuberculosis), thereby distinguishing it from lupus vulgaris. He thought young soft women, lacking energy and vitality, with poor capillary circulation, were especially prone to lupus erythematosus. Seeking to modulate the bodily system, Cazenave recommended mercury and iodides, steam baths and showers, and promotion of sweating.[44] In the 1870s, Hungarian dermatologist Moriz Kaposi was able to specify recurrent systemic accompaniments of lupus, showing that in some cases a severe, generalized febrile eruption occurred, often associated with arthritis, swollen lymph nodes, and weight loss, occasionally resulting in death.[45] At the end of the century, William Osler was confidently writing about the disseminated or systemic features of lupus erythematosus, separating it from discoid lupus, which referred to skin lesions alone. The resources of the modern Johns Hopkins Hospital in Baltimore, Maryland, allowed Osler to recognize a variety of heart, lung, and kidney problems viscerally allied with lupus.[46] Yet diathesis, by then a horribly old-fashioned term, remained as good an explanation as any. After the discovery of the tubercle bacillus, the germ of tuberculosis, some experts hoped this microbe would be isolated from sufferers of lupus, since both constituted disseminated or polymorphic febrile conditions, affecting multiple organs.[47] But again the putative germ frustrated all efforts to find it in persons with lupus.

An unmistakable affliction, childhood diabetes was remarkably rare before the advent of insulin therapy in the 1920s. Rapidly fatal, the disorder usually announced itself with thirst and frequent urination, often associated with a change in mood, the young sufferer becoming quiet and morose before lapsing into coma. The presence of sugar in the urine confirmed diabetes mellitus—blood testing for glucose was not routine until after 1915. During the 1880s, French physicians distinguished two types of diabetes, one rare form in emaciated children and some adults, which led to death within months, and another in overweight older patients who might survive for many years.[48] Later it would become clear that the "thin" type was responsive to insulin.[49] Before pharmaceutical insulin, doctors demanded strict diet and rest, but a sense of hopelessness and despair pervaded the prognosis of "malignant" diabetes. Heredity presumably was responsible for this severe metabolic disorder, combined perhaps with excessive consumption of sugar, though many physicians again could not help wondering if some infection might contribute.[50]

Typical was a boy seven years of age, previously well, who began to tire of playing and to complain of constant thirst. His parents realized he was passing more urine than usual and rapidly losing weight. When a physician in Liverpool, England, saw him in 1885, the child was extremely thin and showed sugar in his urine. "This disease is extremely rare in children," the doctor reported. Within days, symptoms worsened despite a strict diet, coma set in, and the child died.[51] The following year, a doctor on the Isle of Man was called to see a four-year-old boy who was losing weight and wetting his bed. A week later, the child was thinner, mentally confused, incontinent, and vomiting—again with sugar in his urine. He died within days.[52] In 1887, an American physician reported two cases of diabetes in children aged two years: they drank incessantly, passed large quantities of urine, lost flesh, "drooped," gradually "sank," and then died.[53] A physician in Kentucky observed the five-year-old daughter of a German immigrant farmer waste away, drinking water and wetting herself, gradually weakening. A diabetic diet did not prevent her early death.[54] By the end of the century, children with diabetes were common enough not to demand a case report. During the following fifty years, juvenile-onset diabetes mellitus, later named type 1 diabetes, would become, as we shall see, even more prevalent, while remaining for much of this period an etiological conundrum.

Multiple sclerosis presents a revealing example of both the ontological appeal of germ theories and the analytic intractability of some "constitutional" disorders. Described as "paraplegia" in the early nineteenth century, this neurological deterioration was commonly attributed to tabes dorsalis, or spinal cord degeneration, usually attendant on syphilis. An exception was made in the case of Augustus d'Este, whose status as a minor English aristocrat, an illegitimate Hanoverian descendant, may have protected him from such slurs. In 1822, aged twenty-eight, d'Este noticed some indistinctness of vision after attending a funeral. His bilateral optic nerve inflammation returned a few years later, then disappeared, followed by an episode of double vision. Frequently he felt torpid, weak, and numb in his limbs. In his fifties he experienced difficulty walking on rocky ground and descending stairs. Soon he was unable to stand without the aid of a stick. In the years before his early death in 1848 he was confined to a wheelchair. D'Este received a diagnosis of paraplegia, supposedly the result of excessive bile in his system. Accordingly he was regularly bled with leeches, sent off to spas, and even galvanized with electric shocks—but his deterioration continued.[55]

The diagnosis of the German poet Heinrich Heine is more ambiguous. Although Heine and his doctors blamed neurosyphilis or his nervous constitution for his condition, the pattern of recurrence and dissemination in his neurological deterioration is consistent with multiple sclerosis. Long afflicted with severe headaches, Heine woke up one day in 1832 to find two fingers of his left hand paralyzed. The paralysis went into remission then recurred, each time more extensively until eventually it reached his elbow. He suffered episodes of blindness and double vision, giddy turns, along with facial palsies. In 1848 while visiting the Louvre, his legs failed him permanently. Despite leeches, sulfur baths, and copious morphine, the poet died from this aggressive neurological disorder in his fifties.[56]

In 1838, Robert Carswell, professor of pathological anatomy at University College, London, studying autopsies of some cases of paraplegia in which deterioration had been episodic but relentless, observed numerous atrophic patches in the spinal cord, marked by discoloration and softening. Jean Cruveilhier across the Channel noticed these lesions about the same time.[57] A few years later, Charcot consolidated the findings into a new disease entity, which he called "sclèrose en plaques." The Parisian neurologist discerned three key signs of disseminated sclerosis: nystagmus (involuntary eye movements), intention tremor, and scanning of speech—all tending to remit and relapse. He determined that the spinal cord and brain lesions involved loss of the myelin coating of nerves. And he speculated on the relation of the illness to antecedent acute illness—such as typhoid, cholera, and smallpox—or emotional distress.[58] Going further, his student Pierre Marie asserted that the condition must be classed as an infectious disease, though the microbial agent had so far escaped detection.[59] Infection had become the default explanation of disease, but yet again, the specific germ defied discovery.

❋ ❋ ❋

"Why this deliberate, slow-moving malignity?" asked W. N. P. Barbellion in 1917. "Perhaps it is a punishment for the impudence of my desires. I wanted everything so I receive nothing. I am not offering up my life willingly—it is being taken from me piece by piece, while I watch the pilfering with lamentable eyes."[60] An entomologist at the British Natural History Museum, Barbellion—the pen name of Bruce Frederick Cummings—was succumbing rapidly to multiple sclerosis. In the "diary of an intensely egotistical young naturalist," Barbellion, according to H. G. Wells, described the "dark, unforeseen, unforeseeable, and inexplicable fate that has overtaken him."[61] In April 1913, the twenty-three-year old scientist found himself "in a horrible panic—the last few days—I believe I am developing locomotor ataxia [lack of muscle control]. One leg, one arm, and my speech are affected." During the following years, he would suffer intermittent loss of vision, partial paralysis, lack of coordination, fleeting numbness, constipation, and depression. Often he thought of suicide, keeping a pistol handy and seeking out laudanum. "My life seems to have been a wilderness of futile endeavour," the sensitive, ambitious youth wrote despairingly in 1914.[62] Reading Baudelaire, Nietzsche, and Dostoyevsky, he became even more tired, fretful, and morose. In March 1915, he felt particularly self-conscious and introspective. "I never cease to interest myself in the Gothic architecture of my own fantastic soul," he scrawled early in the month. Later, he observed, "I am so steeped in myself—in my moods, vapours, idiosyncrasies, so self-sodden, that I am unable to stand clear of the data, to marshal and classify the multitude of facts and thence draw the deduction what manner of man I am."[63]

Despite the success of microbiology, in the early twentieth century, such deliberate, slow-moving malignity still resisted simple explanation and evaded reduction to any infectious cause. Instead, these conditions seemed better to fit functional models of pathogenesis, evoking a sense of physiological processes disrupted, of bodily integrity undermined from within, of self in peril. Though frequently in remission themselves, concepts of diathesis and predisposition—the medical vocabulary of risk, fate, and destiny—tended to recur in discussions of some chronic, and often systemic, diseases. These mysterious afflictions challenged the sufferer's sense of self or identity, reworking the architecture of the soul.

"My damnable body is slowing killing off all my spirit and buoyancy," Barbellion wrote in November 1915. "Even my mind is becoming blurred." The neurologist Henry Head had diagnosed multiple sclerosis in Barbellion, after briefly considering neurosyphilis, and prescribed rest in the country, along with arsenic and strychnine to help in the bad patches. Barbellion blamed the "bacteria" in his spine. "However sclerotic my nerve tissue, I feel as flaccid as jelly," he wrote. "My God! How I loathe the prospect of death."[64] This self-cultivated young man came reluctantly to believe life a lottery. March 1917 proved especially trying. "All my nerves were frozen, my heart congealed. I had no love for anyone . . . no emotion of any sort. It was a catalepsy of the spirit harder to bear than fever or pain." Bedridden, he was wasting away. His journal soon terminated:

> October 14 to 20. Miserable.
> October 21. Self-disgust.[65]

CHAPTER TWO

Immunological Thought Styles

❋ ❋ ❋

In 1896 the eminently antiseptic surgeon Joseph Lister addressed the British Association for the Advancement of Science, assembled in the mercantile city of Liverpool. "The practice of medicine in every department," he assured delegates, "is becoming more and more based on science as distinguished from empiricism." He urged the attentive researchers to consider the recent studies of Louis Pasteur on alcoholic fermentation, which demonstrated that "all true fermentations, including putrefaction, are caused by growth of micro-organisms."[1] In Berlin, Robert Koch's method of plate culture of bacteria had allowed the isolation of the germs of diseases like cholera and tuberculosis.[2] The eminent surgeon's own introduction of carbolic acid into the operating room had eliminated microbial contaminants and therefore assisted wound healing. Yet, more extensive scientific investigation of measures to counter pathogenic organisms demanded attention. In Paris, Pasteur was attempting to weaken the microbes and then inoculate susceptible individuals with the attenuated form in order to confer on the human frame a sort of immunity. Already he had injected rabbits with brain tissue of rabid animals, producing infection; dried the rabbit spinal cord, thus weakening the rabies virus; and inoculated some of those exposed to the disease organism with this protective substance. Since the procedure seemed to mimic the mysterious defensive mechanism of cowpox, or vaccinia, against smallpox, which Edward Jenner had observed almost a century earlier, Pasteur called his trials "vaccination."[3] Meanwhile, others were trying to produce chemical substances that would combat and neutralize bacterial toxins, or find drugs that might destroy directly the pathogenic microbes within the body. The study of human defense against disease, and the means to enhance immunity, concluded Lister, was an exciting new frontier awaiting further scientific conquest.[4]

A few years later, Lister heard German pathologist Rudolf Virchow echo those concluding remarks. Virchow, who distrusted any facile bacteriological ferment, came to London in 1898 to deliver the Harvey Lecture on Disease, Life and Man. He believed it futile to view microbes as isolated from conditions that favored their growth and spread, and to separate their contribution to disease causation from the resisting powers, or biological individuality, of the organism they attacked. It annoyed him that disease, a complex biological phenomenon, was so readily reduced to infection. It frustrated him, too, that the "secret of immunity" to germs was not yet revealed. The fundamental mechanism of human immunity remained a "great problem whose solution the whole world is awaiting with anxious impatience."[5] How, Virchow wondered, might we strengthen cells in their fight against bacteria? What did it really mean for a body to be immune to a disease? At the *fin de siècle*, immunity evidently exerted a special attraction for scientists and other intellectuals, both as a practical matter and as a philosophic issue. In *The War of the Worlds*, published in 1898, H. G. Wells observed: "By virtue of this natural selection of our kind we have developed resisting power; to no germs do we succumb without a struggle, and to many—those that cause putrefaction in dead matter, for instance—our living frames are altogether immune."[6] Not so the poor Martians who, unaccustomed to the earth's bacterial ecology, met a sudden microbial demise in his popular book. What, then, Virchow and Wells asked, could be the basis of our human immunological distinction?

Immunity was acquiring precise and specific scientific meanings in the late nineteenth century. Once used to designate exemption from political obligations or responsibilities, thereby distinguishing individuals within a community, the term soon came to imply a set of biological contrivances defending the body against particular microorganisms.[7] Previously, physicians had talked generally of resistances or susceptibilities to stimuli that might challenge constitutional stability and health. They sought holistically to restore balance and integrity to the bodily systems of their patients, to aid the healing power of nature. From the 1880s, however, scientists began thinking immunologically. Now they tried to determine precisely the body's defense mechanisms against specific invading microbes. Pasteur had proposed a vague nutritive model for immunity, suggesting that once the special substrate necessary to "culture" the germ within the body—the human test tube—was consumed or depleted, the host became immune to the foreign organism. Unconvinced, his zoologist colleague Élie Metchnikoff argued that circulating cells, which he called phagocytes, could swallow and digest the invaders, thus destroying the parasites. "When one accepts the concept that phagocytes fight directly against pathogens," he wrote, "it becomes understandable that inflammation is a defense mechanism against bacterial invasion."[8] But Koch and his followers, having observed that blood serum, which is free of cells, could transfer this defensive property, insisted that specific immunity must derive instead from humoral elements, circulating chemical substances they called "antibodies," not vigilant phagocytes or depleted nutrients. Whether the mechanism was cellular or humoral, immunity was believed to disaggregate and become specific in an internal struggle for existence of cells and substances within the body, a very particular defensive contest against very particular invaders. To examine this mystery, a new field of scientific study opened up.

It must have been an exciting time to study immunology. Yet, in 1915, when asked to join the editorial board of the new *Journal of Immunology*, the pioneering disease ecologist Theobald Smith told its editor, "Immunology is dead."[9] How could an investigatory enterprise that offered so much promise have come within twenty years to appear so unresponsive intellectually that one of its keenest observers thought it passé? What went wrong? We should begin by taking the pulse of immunology in its youth.

⁂ ⁂ ⁂

"The work of Pasteur and Koch," Paul Ehrlich told a London audience in 1900, "afforded the first basis on which the study of artificial immunity could be . . . undertaken." After graduating in medicine, Ehrlich became expert in using the new aniline dyes to reveal tissue components and structure. Soon he was conducting research in Koch's institute in Berlin on the nature of humoral immunity, trying to determine the character of the soluble substances in the blood that combated microbial invaders, to establish the chemical basis of the function of these antibodies. The discovery by Emil von Behring and Shibasaburo Kitasato in 1890 of some capability of blood to protect specifically against the toxins of diphtheria and tetanus germs presented a pointed biochemical challenge to Ehrlich. Transferred to other animals, these antitoxins could confer protection on them, too, but the composition and mode of action of such antibodies remained a puzzle. "The theoretical explanations of all these facts lagged far behind their practical effects," Ehrlich commented. "I laboured for years trying to shed some light onto the darkness that shrouded the subject."[10]

Through tireless laboratory experimentation, Ehrlich eventually developed his "side-chain" theory of antibody action. During the 1890s, he repeatedly examined and assayed the effects of toxins and antitoxins in animal tissues suspended in test tubes. "Each test tube," he noted, "represents as it were a research animal, uniform in any one series, and one that can be reproduced at will."[11] It seemed to him necessary to differentiate the activities of the toxin: he recognized a "toxophore" element, which caused damage to the animal, and a "haptophore," which conditioned union with the antitoxin. According to Ehrlich, antitoxins or antibodies exist naturally in the body's cell membranes as receptors, and these side-chains showed high affinity for specific foreign substances, incitants called antigens. Ehrlich theorized that once the antigen docks in the matching side-chain, the cell produces an excess of the specific side-chains. As antibodies they circulate in the blood, fastening securely and irreversibly onto the same toxin or microbe (the antigen) wherever they find it. Another mysterious protective substance in the blood—which Ehrlich dubbed "complement"—might also attach tightly and promiscuously to the complex, sometimes causing agglutination or clumping.[12] Ehrlich thus understood immunity in terms of specific affinities and irreversible chemical union. It was just a matter of physical chemistry.

Others shunned Ehrlich's typological and teleological approach. Jules Bordet, a Belgian working in Metchnikoff's laboratory at the Institut Pasteur in Paris, proposed a weaker and more variable interaction of antigen and antibody, one resembling the delicate and gradual adsorption of dyes. He described the immune response as a colloidal process in which one substance disperses through another, not a precise stereo-chemical reaction—more a matter of avidity than affinity.[13] The Swedish physical chemist Svante Arrhenius also argued that antibodies bind only loosely to foreign material and that their combination is reversible.[14] Assuming a virtually limitless set of potential antigens and a restricted number of possible receptors for them, many scientists came to doubt Ehrlich's claims for absolute specificity. How could one person produce an infinite number of specific antibodies? Presumably the immunological repertoire must be more biologically flexible and accommodating.

In Vienna, young immunologist Karl Landsteiner was among those who thought Ehrlich's assertion of strict specificity implausible. Like his mentor Max von Gruber—who had studied in Munich with the iconoclastic hygienist Max von Pettenkoffer—Landsteiner was interested in the variation and transformation of natural phenomena. Concerned with biological gradation and contingency, he was inclined to see antibodies as an expression of biological individuality, not merely as part of a chemical lock and key mechanism.[15] At the turn of the century, Landsteiner was investigating differences in the blood of individuals of the same species, in the process discovering and defining the major human blood groups.[16] For him, *specificity* encompassed a series of smooth quantitative transitions, a summation of numerous nonspecific reactions rather than a rigid and absolute separation of types. No surprise, then, that Landsteiner came, for a time, to favor Bordet's colloidal interpretation of the immune response. Later, after returning to structural chemistry, he would continue to emphasize gradations in reactivity and patterns of cross-reactivity. If an antibody reacted with a variety of related antigens, with different grades of affinity, then fewer antigen receptors were required. This theory seemed to him more biologically realistic—even if more complex and contingent—than anything Ehrlich had claimed.

Landsteiner's commitment to the immune response as an expression of individuality, and his dislike of teleological arguments, led him to take an interest in the obscure disease paroxysmal cold hemoglobinuria. Known since the 1870s, the disease causes red blood cells to break open and release hemoglobin when the body is rewarmed after chilling, leading to bouts of bloody urine, and occasionally anemia. The few medical doctors who had encountered the condition had regarded it as a nervous problem.[17] In 1904, physician Julius Donath approached Landsteiner and asked him to investigate three patients suffering from the disorder. The immunologist isolated some of their red blood cells and placed them in iced water; then he added some of their warm blood plasma, causing the liquid to become colored, as hemoglobin was released from the cells. Landsteiner deduced that the patient's plasma contained the agent that destroyed the red blood cells. He proposed that the normal blood cells acted as antigens—like foreign substances—combining with antibody in the cold and forming a complex that bound to complement at higher temperatures, resulting in cellular disintegration. Apparently, the individual's immune system was paradoxically attacking the body's own cells, thereby challenging teleological explanations predicated on the purely defensive role of immunity.[18] Ehrlich had speculated that, while there was no reason the body would not occasionally produce antibodies against its own tissues—"autoantibodies"—some regulatory mechanism must prevent these substances from acting. The body possessed, he thought, a "horror autotoxicus."[19] In his cautious way, Landsteiner was questioning this assumption.

Around this time, Landsteiner also became absorbed in studying the Wassermann reaction, the new immunological means of diagnosing syphilis infection. Soon after the discovery of the microbial cause of syphilis—the spirochete—August von Wassermann developed, in Koch's Berlin institute, a supposedly specific test for the germ, which revealed the microbe's additional responsibility for chronic neurological diseases, including tabes dorsalis and "general paralysis of the insane." In 1906, Wassermann announced that he could detect antibodies to the spirochete by taking a suspected sufferer's blood serum sample and introducing it to extracts of the organs of humans known to be syphilitic, and therefore presumably containing the spirochete antigen. If specific antibodies against syphilis were present in the patient's blood, an antibody-antigen amalgam would form, attracting complement, which fixed onto the complexes. Then red blood cells were added to the mixture, as an indicator, along with antibodies against them. If there was any complement left, these cells would be lysed, or broken up; but if immune complexes already had used up all the complement, the cells remained intact, a positive result indicating complement fixation.[20]

Distrusting assumptions of absolute specificity, Landsteiner conducted a few experiments using alternative antigens. Curiously, effective antigenic material did not need to come from someone known to be suffering with syphilis. Indeed, the disease was most readily detected when the antigen was normal bovine heart tissue, called cardiolipin, not the spirochete.[21] It was a puzzle. What exactly could the antibodies be attacking? What was consuming them? Avoiding these questions, Landsteiner discreetly moved on to other problems in immunology, but his observations later prompted Edmund Weil, the professor of hygiene in Prague, a satellite city of Vienna, to suggest that the diagnostic antibodies were directed, in fact, against decomposed tissue products. The Wassermann reaction would therefore indicate both syphilitic and other damage to organs, and not primarily detect the presence of the foreign microorganism. Rather than a specific test for syphilis, then, perhaps the Wassermann reaction was demonstrating an immune response to the body's own damaged tissues.[22] Weil's further study of this strange phenomenon was prevented by his untimely death.

Before World War I, a commitment to the study of biological individuality and sensitivity distinguished much of the medical culture of Vienna.[23] Landsteiner was one of a group of physicians there who criticized the biochemical reductionism emerging from Berlin and who chose instead to investigate variability and contingency in immune responses. For a time he joined the circle around Gruber, the discoverer of the immunological basis of blood agglutination and an opponent of Ehrlich's facile emphasis on fixed species and standardized specificities. Gruber and Landsteiner continued to resist simple disease ontology, preferring a more physiological view of the immune response. They often discussed these matters with Rudolf Kraus, an eccentric serologist who introduced them to pediatricians Clemens von Pirquet and Béla Schick, and to his friend Wassermann, who frequently visited family in Vienna.[24] These cultured physicians mingled in Vienna's cafés and salons with the artists, musicians, philosophers, and architects who made the imperial city a "hothouse of modernity," according to historian Carl E. Schorske. "Narcissism and a hypertrophy of the life of feeling were the consequence" of such intellectual mixtures, he writes, giving rise to a "culture of sensitive nerves," a focus on multiplicity and indeterminacy, and a suspension of absolutist claims.[25] But rising anti-Semitism and the beginning of war soon imperiled the modern thought style in which they were participating, eventually leaving the survivors of the group as isolates in the German immunological mainstream.

❋ ❋ ❋

Several physicians in Paris also maintained an interest in the variability of the immune response during this period, even though their institutional milieu was unsympathetic. Generally, scientists at the Institut Pasteur preferred to concentrate on developing practical solutions, such as vaccines and immune sera, to disease problems, avoiding arcane theorizing. Refusing to conform, Metchnikoff continued to speculate on the digestive functions of phagocytes and Darwinian conflict within the body. Drawn to contemporary theories of autointoxication, he also believed that toxic products from normal digestive processes caused disease and early death.[26] But the preoccupations of the Russian zoologist increasingly seemed quirky and unproductive to most other Pastorians. Meanwhile, beyond the reach of the Institut Pasteur, the physician and physiologist Charles Richet was fervently investigating individual variation in the reaction of animals to injections of toxin, wondering how difference in constitution and temperament might influence their responses.

Once a student of the physiologist Claude Bernard, Richet maintained a broadly biological understanding of bodily function, whether normal or pathological. Like most investigators interested in biological individuality in this period, he chose humans, dogs, or cats as research subjects, rather than the readily reproducible rodents favored in Berlin. In 1902, he tried to immunize his dogs with toxin from sea anemones: those surviving received a second injection, but some of them soon had trouble breathing and died suddenly. Richet and his colleague Paul Portier discerned a difference between the initial, toxic, dose and the smaller, sensitizing dose, which resulted in a reaction they called anaphylaxis. It seemed anaphylaxis and immunity were closely linked: both were specific, developed after a lag period, and could be transferred passively through cell-free blood serum from the affected animal to another. Anaphylaxis, then, must be a manifestation of the immune system's extreme sensitivity to a foreign substance. It represented, in this sense, the risk of immunity, an alarmist overreaction.[27] Most appealing for Richet was the idiosyncrasy of anaphylaxis. While some animals became sensitized, most did not. Thus, it reflected individual differences in immunological character or susceptibility. Devoted to socialism and eugenics, Richet speculated that anaphylaxis might conveniently select out and destroy weaker individuals, thereby protecting the species.[28]

"All of us at this present moment are, consciously or unconsciously, humoralists," Richet told the delegates to the 1910 International Congress of Physiology in Vienna. "That is to say, we look upon the chemical constitution of our humours as being the basis of all biological phenomena." He went on to describe his research on anaphylaxis and biological individuality. The lesson was clear. "The chemistry of the imponderables becomes necessarily . . . the chemistry of functions," leading to the "physiology of the individual." Although psychological individuality was well known, he had discovered a distinct "humoral individuality." Richet was especially struck by variations in humoral memory and irritability, the aptness "to modify one's self under the influence of the feeblest external actions." This immunological self thus recognizes two fundamental laws, "imponderability and instability."[29] Chemistry might govern the body, but for Richet it was a remarkably idiosyncratic chemistry.

For Clemens von Pirquet and many other Viennese in the audience, Richet's address contained few surprises. In 1902, with his colleague Béla Schick, Pirquet had observed a few children suffer severe reactions to the new antisera, which were derived from the blood of animals (usually horses) exposed to various specific diseases and which therefore contained substances that might confer the corresponding immunity on the patient. One young boy with scarlet fever received an injection of antiserum and seven days later became ill with high fever and generalized itching. Another child was injected with a second dose of diphtheria antiserum some fifty days after taking his first dose: within fifteen minutes he vomited and showed swelling of his face and neck and an itchy rash over his body.[30] It was evident to the young investigators that some individuals could develop systemic reactions to injections of antitoxins and sera, especially if these were repeated. That the illness, which they called serum sickness, took time to emerge suggested an immunological cause, not any direct injury. As Pirquet and Schick later put it, "The conception that the antibodies, which should protect against disease, are also responsible for the disease, sounds at first absurd."[31] Indeed, they wondered if only a few of their Viennese colleagues and a couple of Parisian renegades would countenance the idea.

By 1906, Pirquet had coined a new word, *allergy*, to express "the change in condition that an animal experiences after contact with any organic poison, be it animate or inanimate." Allergy therefore implied changed reactivity or sensitivity. It encompassed a range of symptoms and signs, from minor rashes to severe breathing problems. The provocation might include dust, hay, feathers, animals, food, pharmaceuticals, vaccines, and sera—anything really. "This explanation involved also quite a new conception of an antibody," Pirquet wrote confidently. "Thus far the antibodies were numbered among the protective substances, which is just the contrary of the supposition. . . . The principal new conception consisted in the suggestion that a disease might be due indirectly to an antibody."[32] Initially, the proposition that a regulatory or defensive mechanism gone awry—an alteration in reactivity—might give rise to disease elicited considerable skepticism. Many physicians and scientists held to the belief that the immune system was designed solely for defense. They still preferred to explain hay fever and asthma as nervous conditions. Resentfully, Richet claimed that the term *allergy* was redundant, since he had already called it anaphylaxis. Others asked if the notion differed significantly from older assumptions of diathesis. However, within a decade or so, allergy, like psychoneurosis, was everywhere, the epitome of modern malaise.[33]

Conventionally, the irresistible rise of germ theories in the late nineteenth century is seen as marking the triumph of ontological concepts of disease, with the establishment of tight and specific bonds between cause, effect, and response. In this Linnean taxonomy, each species of germ is supposed to incite one type of disease and elicit perfectly matched antibodies. George Bernard Shaw captured the moment in a remark in 1911: "We are left in the hands of the generations which, having heard of microbes much as St Thomas Aquinas heard of angels, suddenly concluded that the whole art of healing could be summed up in the formula: Find the microbe and kill it."[34] Undoubtedly, the conceptual framework in which disease is located was changing rapidly during this period, yet still we find reflective physicians and scientists with more physiological or integrative inclinations thinking about the biological complexity of human responses to disease and being concerned with idiosyncrasy and variation—with modern diatheses, in a sense. This somewhat marginal cohort tended to interpret immune responses in terms of delicately graded reactivity and sensibility, rather than narrow specificities and facile teleology. Some of them postulated forms of heightened or pathological reactivity, claiming that antibodies might attack the body's own tissues, causing disease, or retaliate excessively against external agents, bringing on allergy. In each case, the normal defensive processes of the body could become pathogenic and dysteleological—just like fever.

If idiosyncrasy and variation in reactivity appealed to many of these early immunologists, they constituted just one part of the broader revival of interest in biological and psychological individuality taking place in the early twentieth century. Medical historian Owsei Temkin observed that ontology, the classification of separable entities, creeps back into even the most rigorously physiological frameworks of disease; yet the contrary is also true, with the ineluctable return of repressed physiological or integrative explanations.[35] Thus, we find inquiries into individuality—or at least the systemic conditions of abnormality—surfacing in multiple fields related to medicine and psychology during this period, not just in immunology. In 1902, the London physician Archibald Garrod, for example, found time between studies of the etiology of rheumatoid arthritis to investigate disorders of metabolism and speculate on "chemical individuality."[36] Later, the English physiologist Ernest Starling defined "hormones" as chemical messengers carried through the bloodstream from one organ to another, the "fundamental means for the integration of the functions of the body." Moreover, he believed that disorders of the special internal glands that secreted these hormones could lead to various ailments.[37] In 1911, the Polish biochemist Casimir Funk isolated the active ingredient of brown rice that prevented beriberi, calling it a "vital amine" or "vitamine": he claimed that many other constitutional disorders resulted from nutritional deficiencies or imbalances causing abnormal metabolism.[38] In the Vienna Psychoanalytic

Society during this period, Alfred Adler was strenuously promoting his theory of "individual psychology," arguing for viewing the individual holistically, not reductively, and adopting the words *allergy* and *anaphylaxis*.[39]

Those urging attention to individuality and organic identity evidently comprised a diverse group, but they shared an aversion to exaggerated reductive approaches and clinical routinism. Their understanding of normal function and pathological agency was complex and nuanced. The onset of World War I appears for a time to have discouraged many of these enthusiasts, cutting short or muffling most medical analysts of individual reactivity and sensibility and rechanneling their interests, though never quite extinguishing them. Perhaps dismemberment and mutilation of bodies in the trenches and the wounds that the war machine inflicted on flesh and mind made it hard for a while to cherish individuality and human distinction. Not until the late 1920s does one find renewed curiosity toward biological individuality. In the interwar period, the Polish immunologist and hygienist Ludwik Hirszfeld, who with his wife, Hanna, had demonstrated ethnic variations in the distribution of Landsteiner's ABO blood groups, was developing his "constitutional serology," claiming that blood groups overlapped with hereditary predispositions to certain diseases. Serology, he believed, might reveal the constitutional basis of disease susceptibility. But the idea that natural antibodies defined biological individuality, especially pathological proclivities, did not catch on—not then, anyhow.[40] In America during the 1930s, biochemist Leo Loeb, who before the war studied the action of reproductive hormones on cancer growth, began to wonder about the nature of the "individuality differential," each animal's "specific chemical constitution," which determines whether foreign tissues and organs would prove acceptable to the organism.[41] The problem of individual variation continued to haunt medical science—at least along its margins. Although frequently characterized as reductionist, biomedicine in the twentieth century never fully escaped its constitutional ties.[42]

So, why did Theobald Smith dismiss immunological inquiry in 1915? Only a few years earlier, scientists in Paris and Vienna had investigated individual reactivity, examining idiosyncrasy and sensibility in immune responses, even postulating pathophysiological mechanisms of disease causation that implicated the body's own regulatory system. Surely such biological vision once must have impressed the disease ecologist, suggesting new styles of host-parasite interaction and evolutionary influence. But by 1915 physicochemical studies of antibody-antigen binding had come to dominate immunological research, and not only in Berlin. The Institut Pasteur focused ever more narrowly on vaccine development and inventing diagnostic tests. Meanwhile, clinical allergy was diverging from the immunological mainstream, becoming increasingly concerned with modulating patients' responses to specific antigens and largely discarding broader biological interests. Around 1911, "allergists" in London devised a means of desensitizing sufferers through inoculation with the allergen, thus treating the pollen as a toxin and trying to vaccinate against it. Much of the nascent specialty soon was obsessed with specificity and standardization, largely—though never completely—abandoning physiological inquiry in favor of ontological reasoning and the attendant seductive clinical routine.[43] No wonder, then, that Smith imagined he was witnessing the end of immunology as a major field of intellectual endeavor, as a means of addressing important biological questions.

❋ ❋ ❋

Increasingly isolated in Vienna, a victim of postwar privations and anti-Semitism, Landsteiner decided to leave the declining city. After a brief sojourn in the Netherlands, the immunologist moved to the Rockefeller Institute in New York City, a thriving medical research center on the east side of Manhattan. He was joining an institute that boasted physiologist Jacques Loeb, the model for Max Gottlieb in Sinclair Lewis's novel *Arrowsmith*, and the French surgeon Alexis Carrel, studying methods of organ transplantation and tissue culture.[44] Although Landsteiner's career linked Vienna and

New York, the transmission of ideas and culture across the Atlantic was patchy at best. Now stiff and formal, well into aloof middle age, the immunologist exerted little direct influence on his institute colleagues, even those inclined to view disease through a biological lens and work on similar problems. "He is an astonishing person," wrote allergist Arthur F. Coca, "and a great man in our subject, and I am sorry that not everybody at the Rockefeller seems to appreciate him."[45] Until his death in 1943, Landsteiner continued steadily to conduct research in New York on blood groups, allergic reactions, sensitization, and specificity, avoiding extensive exchanges with colleagues.

Established in 1910, the hospital of the Rockefeller Institute gave prominence to clinical research, bringing science to the patient's bedside. The first director, Rufus Cole, came from the Johns Hopkins Medical School, where he had served under physician Lewellys F. Barker and conducted bacteriological experiments. Cole resolutely regarded Hopkins as the model for the Rockefeller Institute, and he insisted on recruiting full-time physician-scientists to the hospital, just as they did in Baltimore, and giving them control of a ward and time for research. He decided the hospital should concentrate on five major medical challenges: poliomyelitis, syphilis, heart disease, "intestinal infantilism" (celiac disease), and lobar pneumonia, his own specialty.[46] Both hospital and institute were embedded in the cosmopolitan city, which grew rapidly between the wars, despite the Great Depression. Around the middle of the 1930s, New York City was gaining a reputation for exuberance, emerging as a site for modernist experiments in literature and the arts. New skyscrapers, like the Chrysler Building, the Empire State Building, and the Rockefeller Center, had transformed midtown Manhattan; Robert Moses was building highways and rearranging the parks; the city buzzed with baseball, radio, the music of Duke Ellington, and motorcars. New York exerted an edgy metropolitan attraction for scientists from Europe and the American hinterland.

Among the first Rockefeller Hospital recruits, Homer Swift had studied tabes dorsalis, by then recognized as a neurological complication of syphilis. To treat the condition, he developed a preparation of Salvarsan—one of Ehrlich's exciting chemotherapeutic agents, specific for syphilis—that was not irritable to nerve tissues. But in the 1920s, Swift strayed into studies of rheumatic fever—a common yet puzzling condition that could include migratory arthritis or heart inflammation and damage or skin rashes and nodules or any combination of these. Unable to recover a single causative organism from patients, Swift speculated on some sort of allergic process underlying the disease. Since 1900 it had been clear that antecedent acute infection with the germ streptococcus was somehow associated with rheumatic fever, but the microbe was not usually present when the disease actually occurred. In the late 1920s, Swift came to believe that the microorganism could not be directly responsible for tissue damage; rather, he thought the problem was the hypersensitivity of some tissues to a few strains of streptococcus. "The reactivity of the tissues is changed," he wrote, "in such a manner that they respond in a peculiar fashion to certain non-specific forms of insult."[47] That is, rheumatic fever must be the consequence of the idiosyncratic responses of the host, not the virulence of the germ.

Swift's cautious interest in the contribution of allergy to disease causation was not unique. At the same time, Harvard microbiologist Hans Zinsser also was arguing that chronic disease might derive from hypersensitivity of tissues to certain bacteria, an allergy or overreaction to germs. Zinsser regarded the tuberculin reaction, the delayed local inflammation at the site of an injection of killed tubercle bacilli, which cause tuberculosis, as a typical example of bacterial allergy. This puzzling tissue reaction, evidently dependent on hereditary predisposition, seemed unrelated to antibody production. Always interested in "the philosophical implications of scientific fact," Zinsser found the process intriguing.[48] He supposed that the "fundamental biological significance of bacterial allergy is an increased specific adjustment of the tissues for response to the stimulus of infection."[49] Although he considered this "increased capacity to react" to be mostly protective, he feared it could sometimes "result in pathological changes and disease."[50]

All the same, study of such variation in human reactivity remained a minority interest.

In a Rockefeller Institute laboratory adjacent to Swift, Thomas M. Rivers, a Hopkins graduate, was investigating a number of putatively viral diseases. A young Methodist from Georgia with a waspish tongue, Rivers had spent much of his youth as a clinical assistant in Panama, becoming familiar with tuberculosis and venereal disease, fiddling with Wassermann tests, and dispensing Salvarsan. During World War I he studied pneumonia and influenza, developing a special interest in susceptibility and resistance to these scourges. Appointed to the Rockefeller Institute the same year as Landsteiner, the pugnacious Rivers found the émigré a "hard customer," industrious and evasive. Swift and his colleagues proved more amiable and engaging. Rivers found their ideas about the allergic basis of many diseases especially intriguing.[51] They got him thinking about the brain inflammation, or encephalitis, recently recognized as a rare consequence of vaccination against smallpox; he recalled, too, a few cases of encephalitis in which the myelin coating of the nerves had broken down after early uses of the Pasteur rabies vaccine. Neither phenomenon could be attributed to the direct action of any virus on the brain; perhaps they represented an allergic response. Rivers decided to inject some brain tissue into monkeys and see what happened.

Starting in 1932, Rivers and his colleague Francis Schwentker injected twice each week an emulsion of normal rabbit brain into their rhesus monkeys. After six months, Schwentker came to Rivers and said, "Tom, there is something the matter with those monkeys." They watched the animals closely. "Sure enough," Rivers recalled, "one or two of the monkeys were slightly ataxic [unsteady] and looked kind of seedy. A day or two later they began to tremble and became weak in the legs, and very shortly thereafter they became so ataxic that they couldn't get around the cage. At this point I killed them." Examining their brains at necropsy, the scientists found to their "astonishment" evidence of demyelinating encephalitis—injection of the brain emulsion apparently had led the recipient body to attack its own brain tissue, though the investigators resisted calling the reaction an autoimmune process.[52] In any case, they had discovered what came to be called "experimental allergic encephalomyelitis," a model for understanding multiple sclerosis, or demyelinating disease, in humans—the affliction that had killed Barbellion a few years earlier. All the same, according to Rivers, other scientists continued to object that

> allergy is being too extensively invoked in efforts to explain the manner in which a variety of abnormal conditions arise, and that it is unnatural for the body to react to invasion in a way that will result in injury to itself. In this connection, however, one must keep in mind the fact that allergy, or altered reactivity, is a universal phenomenon associated with all infections. . . . Furthermore, not all states of allergy are purposely protective or beneficial to the body.[53]

"To me," Rivers asserted more confidently many decades afterwards, "it's a profound biological phenomenon to learn that the tissues of a person or animal can create antibodies that will result in disease or the death of that person or animal."[54]

Meanwhile, Swift's growing obsession with altered reactivity of tissues lured to his laboratory Joseph E. Smadel, a ribald, young Midwesterner who had studied with Leo Loeb in St. Louis. Smadel wanted to find out the circumstances in which antibodies could be destructive as well as protective. He began injecting rabbits with suspension of rat kidney, producing antibodies against kidney tissue; then he took the anti-kidney serum and inoculated rats, causing degeneration of their kidneys. In most animals, their renal disease became chronic.[55] Swift wondered if his junior colleague thus had provoked "a kind of reverse anaphylaxis, in which certain tissue elements unite with antibodies in the immune serum."[56] Normal kidney cells were acting as the antigen to the introduced antibodies. "It is possible," Swift reported, "that there are formed two classes of antibodies, one sensitizing and the other immunizing, the one deleterious and the other beneficial, and that the outcome in some chronic infections may, in part, be determined by the relationship between the two."[57] Allergy and defense seemed to operate together in the body, in constant tension.

Impressed, Schwentker also decided to try to produce anti-kidney antibodies in rabbits. He immunized the animals with rabbit kidney tissue mixed with toxin from the staphylococcus germ. Although complement fixation allowed him to detect opposing antibodies in the rabbits' blood, their kidneys did not seem damaged. Additionally, Schwentker found, using the same method, that several patients in the hospital with scarlet fever, a result of streptococcal infection, happened to produce antibodies against kidney cells. He wondered whether in some individuals the streptococcus toxin functioned as a carrier protein—or *schlepper*, as the Germans used to call it—combining somehow with kidney cells. Together they could serve as an effective antigen, inducing "some type of antibody" that was able to "react on the kidney giving rise to acute hemorrhagic nephritis [inflammation]." Schwentker believed the process to be "anaphylactoid" in character, the manifestation of altered reactivity in some individuals. But then he backed off, just as he and Rivers had done in their earlier studies on encephalomyelitis. Again he resisted using terms like autoantibody or autoimmunization. Instead, he conceded, the "entire theory discussed here is based on only a frail foundation."[58]

This inchoate interest in allergic or immune reactions within the body against its own tissues soon dissipated at the Rockefeller Institute. Smadel moved over to join Rivers, who became distracted as he searched again for new viruses, a more dependable enterprise, and a particularly creditable one in his intellectual community.[59] Swift's research appears simply to have stalled. Landsteiner was preoccupied with blood grouping before his retirement, though he did maintain a few inquiries into sensitization of tissues and individual reactivity. In *The Specificity of Serological Reactions*, the elderly Viennese reviewed the possibility of animals' mounting immune reactions against their own tissues, citing the Wassermann reaction, his old studies of paroxysmal cold hemaglobinuria, and the research of Schwentker and Rivers.[60] While Landsteiner avoided talking with colleagues, evidently he was watching them closely.

❋ ❋ ❋

In the 1930s in Lwów, Poland, the serologist Ludwik Fleck also was thinking about the Wassermann reaction. When free of routine work in the bacteriology laboratory, Fleck tended to reflect on the philosophy of medicine. Initially, in addresses to other intellectual doctors in Lwów, he chose to speculate on the contingency of disease classifications and the complex interactions of host and parasite, focusing on Edmund Weil's notion of latent infection.[61] In 1927 Fleck spent several weeks in Rudolf Kraus's laboratory in Vienna, on the advice of his mentor Rudolf Weigl, another staunch critic of Ehrlich's claims of bacterial specificity and an advocate of colloidal explanations of antigen-antibody interaction. Fleck returned to Poland brimming with ideas about the artfulness of the Wassermann reaction and what it might reveal about the cultural and social conditioning of scientific knowledge. How did a procedure so variable and unyielding to explanation become the conventional scientific test for syphilis infection? What made it a scientific fact?

For Fleck, the Wassermann reaction made sense only within a particular *Denkstil*, or thought style; in this case, it confirmed assumptions about "syphilitic blood," elaborating timeworn associations between the disease and tainted or corrupt blood. Within the *Denkkollektiv*, the thought collective, of bacteriologists and physicians, a positive Wassermann reaction simply implied infection with spirochetes. Fleck argued that such "stylized" perceptions—expressed as dogma or a set of axioms—depended intimately upon social affiliation, on the collective to which one belonged; yet for him, the test had become a complex, subtle, and tacit performance, revealing the host's reaction to infection, and not primarily its etiology. Although commonly interpreted as the specific test for the presence of spirochetes, the Wassermann reaction could thus connote instead a variation in individual reactivity. Like Weil, Fleck believed the test actually detected antibodies directed against damaged tissue, not the pathogenic agent. But the dominant immunological thought collective continued to resist explanations predicated on reactivity and sensibility. "From time to time," wrote Fleck, "very promising relations and vistas open up, only to vanish again like so many mirages."[62]

Published in 1935, Fleck's book *Genesis and Development of a Scientific Fact* sank into obscurity until discovered in the 1970s and recognized as a foundational text in critical science studies. Received in the thought collective of science studies as a philosophical tract, it may be read still more aptly as a poignant argument against the "social constraint on thought" exercised in interwar immunology and medicine. It was, in effect, Fleck's plea for more attention to biological individuality in immunological reasoning. He might have called it "Anti-Ehrlich." But by the 1940s a gradual change in the immunological thought collective's "readiness for directed perception" was occurring.[63] The idea of autoimmune disease was on the verge of seeming reasonable. But the Lwów immunologist did not participate in the shifting thought style. Incarcerated in the extermination camps of Auschwitz and Buchenwald, Ludwik Fleck was struggling to survive.

CHAPTER THREE

A Sense of Unlimited Possibilities

❋ ❋ ❋

"About current scientific speculations there is one characteristic, subtle, perhaps, but profound and far reaching, which distinguishes them from the Victorian age," wrote J. W. N. Sullivan, an English historian of science, in 1921, conjuring up a "spirit." "This spirit is chiefly a sense of unlimited possibilities, a sense that the radically new and unprecedented may be upon us."[1] Credited with explaining the physics of relativity and quantum mechanics to the British public, Sullivan eagerly propagated the latest scientific discoveries, and their presumed idealist implications, through London literary salons between the wars. Aldous Huxley and T. S. Eliot learned much of their science from him; the occultist Aleister Crowley listened as he discoursed on Albert Einstein and Niels Bohr. Born and bred in the East End of London, though claiming genteel Irish origins and connections to James Joyce, Sullivan adapted rapidly to Bloomsbury after World War I, assisting J. Middleton Murry with editorial work for the *Athenaeum*. But some literary figures resisted his scientific enthusiasm. With a sharp eye for the *parvenu*, Virginia Woolf regarded him as "too much of the India-rubber faced, mobile lipped, unshaven, uncombed, black, uncompromising, suspicious, powerful man of genius in Hampstead type for my taste."[2] D. H. Lawrence dismissed his abstract scientific verities. And then, at the height of his influence in the early 1930s, Sullivan developed a "creeping paralysis," like Barbellion's disease. To his distress, he learned there was no agreed scientific explanation of his multiple sclerosis, and no effective treatment. He died in 1937, aged fifty-one, shunned by most of his former literary associates. But W. J. Turner, the editor of the London *Daily Herald*, remembered Sullivan as "one of the most loveable, brilliant, and uncomplaining of men, who lived and died with an heroic fortitude that equaled his gusto and love of life."[3]

Ten years or so after Sullivan's death, a plausible scientific explanation of the cause of multiple sclerosis did emerge, though it was not radically new and unprecedented. In the 1940s, a number of scientists reverted to the study of experimental encephalitis, or induced brain inflammation, regarding it as a model for other demyelinating diseases, such as multiple sclerosis, and they began to explore more confidently the contribution of antibodies directed against the body's own brain tissue. The idea of allergy, or the enhanced reactivity of the immune system, retained an appeal for many investigators, though eventually they would come to describe the mechanism as an "autoimmune" process, the result of a misguided and erroneous antibody reaction against a "self" antigen. Research into other diseases of disputed or unknown cause also turned up, often unexpectedly, some hints of autoimmune activity. Soon after World War II, clinical investigators in North America and Europe uncovered the autoimmune origins of systemic lupus erythematosus and rheumatoid arthritis. In the 1950s, various blood disorders and chronic inflammations of thyroid, liver, gut, and kidney also would secure an autoimmune explanation, displacing conjectures about infection or hereditary predisposition and diathesis. Once an immunological solecism, autoimmunity became widely available as a conception of disease causation, a candidate explanation as credible as any other. Although the idea of autoimmunity, as we have seen, was hardly unprecedented, its accumulation of plausibility, its stabilization, in the postwar years marks a change in contemporary assumptions about the normal human body and its pathologies, as well as a shift in theories of biological individuality and the nature of the self.

Significant postwar investments in clinical research made possible the discovery of the various autoimmune diseases. Indeed, autoimmunity was imaginable and provable only within a distinctive institutional matrix.[4] Not only was financial and institutional support for medical research greatly expanded during the 1940s and subsequent Cold War; the links between laboratory and clinic became more dense, intricate, and productive. Therefore, it became more common for a hospital patient to be treated also as a research subject, a conjunction that sometimes gave rise to revealing, and even surprising, results. Additionally, advances in technique could make medical research far more effective and gainful than before. The use of "adjuvants"—special enhancing agents—amplified the effects of immunization with normal tissues, making laboratory experiments much less laborious than those Tom Rivers and his colleagues had conducted. Tests for detecting the liaison of antigen and antibody were refined during this period, and their scope expanded, thereby exposing many more autoimmune processes. As medical investigation enlarged its range and capacities, developing more efficient and precise experimental procedures and ramifying into hospital wards, the category of autoimmune disease would grow and recruit apace.

❋ ❋ ❋

The cause of multiple sclerosis still engendered considerable scientific debate well into the 1940s. One group of investigators continued to claim that the disease was degenerative, the consequence of toxins or other environmental factors colluding with a hereditary diathesis, while another persistently argued for an inflammatory condition, localized as acute lesions of nerve tissue. The reasons for the sporadic destruction of the myelin sheaths of nerves remained baffling, especially since no bacteria, viruses, or toxins could be consistently isolated in persons with multiple sclerosis.[5] Reflecting on the rare occurrence of demyelination after rabies vaccination, some scientists and clinicians in the 1930s and 1940s joined Rivers in wondering if multiple sclerosis might be a manifestation of an allergic reaction.[6]

The possibility of an allergic mechanism in multiple sclerosis intrigued Armando Ferraro, a neuropathologist at the New York State Psychiatric Institute and an Italian liberal who had fled the dictatorship of Benito Mussolini. "Is there in the neuropathological features of demyelinating processes," he asked in 1944, "any analogy to or identity with the pathologic features described in the brain in experimental allergic states?"[7] Ferraro repeated Rivers's experiments, producing anti-brain antibodies and demyelination of the central nervous system of monkeys after injecting an emulsion of rabbit brain into their muscles. Moreover, he recently had examined the brains of two people who had died from encephalitis following scarlet fever, and he found evidence of an allergic reaction against neural tissue.[8] In 1944, Ferraro decided to compare the pathological pictures of cerebral anaphylaxis, induced through injection of egg white into the brains of monkeys, with demyelinating disease. He observed the same pattern at autopsy, implying that an immunological process—an allergic reaction, in his words—was responsible for both kinds of pathology. But what prompted this reaction? What was the natural antigen or incitant? Ferraro speculated on diet, bacterial toxins, and other external agents as contributing factors. He even wondered if the myelin sheath itself could constitute a provocation and target for antibodies.[9]

Ferraro was careful not to challenge directly the etiological theories of Tracy J. Putnam, the director of the Neurological Institute at Columbia University and probably the leading New York neurologist at the time.[10] Emphasizing nonimmunological causation, Putnam believed that obstruction of small veins could result in loss of myelin from nerve cells, thus causing multiple sclerosis. He claimed that the blood was more prone to clotting in cases of multiple sclerosis, generating thrombi that blocked the vessels supplying nervous tissue. But Putnam and colleagues experienced difficulty in explaining how the rest of the nerve was preserved, even as the myelin degenerated.[11] Still, Ferraro cautiously concluded that the vascular problems championed by Putnam might develop as a process secondary to the allergic reaction he described.[12] The neurologist, however, suggested that the brain emulsions Rivers and Ferraro injected into monkeys acted primarily as coagulants, not antigens.[13] All the same, Putnam made sure to appoint an immunologist, Elvin A. Kabat, to the Neurological Institute to investigate any possible allergic involvement.

Both the Psychiatric Institute, where Ferraro was chief neuropathologist, and the Neurological Institute, Putnam's domain, were integrated into the Columbia-Presbyterian Medical Center in Washington Heights, New York. Formed with Rockefeller Foundation support in 1929 as the first purpose-built academic medical center, Columbia-Presbyterian linked the medical school with hospital wards and research institutes, all located together uptown on the old Highlanders baseball property.[14] The academic medical center promoted collaboration between laboratory researchers and clinicians, cultivated a "research mindedness" in medical students and residents, and ensured that science could readily be applied at the patient's bedside. A range of clinical conditions became available for laboratory investigation, which might involve biochemical testing, physiological and pathological studies, or modeling with experimental animals. In academic medical centers, those vast and complicated urban structures, every interesting case turned into a potential research subject.[15]

In 1940, Putnam approached the biochemist Michael Heidelberger, a colleague nearby at the College of Physicians and Surgeons, Columbia University, asking if Heidelberger could find a young researcher skilled in analytic chemical methods for measuring antigens and antibodies. Elvin Kabat, the appointed scientist, recalled that Putnam told him that "he understood I had lots of my own ideas and that I could work on whatever I wished except he hoped I would not discriminate against neurological problems."[16] Known as a headstrong radical, Kabat had graduated in science from City College, New York, during the Depression, supporting his family by toiling in Heidelberger's laboratory. Studying the physico-chemical properties of purified antibodies with Heidelberger and then with Arne Tiselius at Uppsala, Kabat discovered that circulating antibodies came from the gamma globulin fraction of blood protein. During the 1930s, the young scientist was flirting with the Communist Party, traveling to the Soviet Union, and visiting the International Brigade in the Spanish Civil War to express his solidarity with the struggle against fascism. Such political commitment was not unpopular among his cohort, but it distinguished him from the older generation of researchers, most of whom boasted medical degrees, middle-class origins, and moderately liberal inclinations—and during the Cold War it would prove an enthusiasm disruptive to his career.

Expediently, Kabat decided at the Neurological Institute to focus his research on multiple sclerosis. Examining the cerebrospinal fluid of patients with the disease, he detected high levels of gamma globulin, implying some immunological activity in the brain. In order to establish the exact significance of this finding, he began to concentrate on the biochemical and quantitative aspects of allergic reactions. At the end of the war, Kabat, along with the neuropathologist Abner Wolf and Ada E. Bezer set about repeating the experiments of Rivers and Schwentker, trying to produce acute disseminated encephalomyelitis in monkeys, presuming it would be a model for multiple sclerosis. The original experiments required many injections of brain emulsion, and the disease took a long time to appear. But Jules Freund, a bacteriologist also laboring in New York City at the time, had recently demonstrated that adding an emulsion of paraffin oil and killed mycobacterium, the cause of tuberculosis, to the antigen caused a prolonged and more immediate immune reaction. Using Freund's adjuvant as a sort of reinforcement with normal rabbit brain tissue, Kabat found that it took only a few injections to cause encephalitis in rhesus monkeys and confirm the disease's allergic mechanism. The scientists observed with fascination the relapsing and remitting course of the nerve damage in their animals: a blind monkey, for example, would recover its vision only to lose it again a few days later. The brain tissues, once fixed, stained, and examined microscopically, showed degeneration of the myelin coating of nerve fibers, a pathological picture similar to that of multiple sclerosis.[17] Soon after completing the experiments, Kabat heard from Isabel Morgan, a virologist at Johns Hopkins, that she was injecting monkey spinal cord tissue containing poliomyelitis virus with Freund's adjuvant into other monkeys and obtaining the same result.[18] Both investigators suggested that an immunological process gone awry would explain experimental allergic encephalomyelitis and, accordingly, multiple sclerosis. As Kabat and his colleagues put it in 1947, "autoimmunization as a possible mechanism in the production of comparable lesions in man must be considered."[19] Soon after, the new National Multiple Sclerosis Society awarded Kabat its first research grant.

The use of Freund's adjuvant transformed immunological investigation after World War II. A Hungarian bacteriologist and hygienist, Freund moved to the United States early in the 1920s, visiting the Rockefeller Institute before working at the University of Pennsylvania and then Cornell. Intrigued by immunological sensitivity to tissue injury and infection and curious about the Wasserman test, Freund, like so many Austro-Hungarian medical graduates, wanted to study anaphylaxis and allergy. But the exigencies of making a living in America between the wars meant he often undertook routine analysis of tuberculosis cases and labored for years producing biological products. In the early 1940s, while directing the laboratories of the New York City Health Department, Freund became interested in the heightened immunological reactivity of animals infected with tuberculosis. Tuberculous guinea pigs, for example, produced more antibodies against other introduced antigens than their non-tuberculous counterparts.[20] With Katherine McDermott, Freund mixed tubercle bacilli with paraffin oil and added some horse serum, then injected guinea pigs with the concoction. Compared to other guinea pigs receiving only horse serum, those subjected additionally to the adjuvant—the auxiliary oil and mycobacteria mixture—showed an amplified and prolonged response to the injections. The adjuvant greatly enhanced the reaction of the immune system against the introduced antigen.[21] The explanation for this sensitization remained obscure, but no one doubted the usefulness of the enigmatic technique in accelerating immunological experiments. Within a year or two of its invention, the adjuvant spread beyond New York, becoming a mainstay of immunological laboratories around the world. As a leading immunologist reflected in the 1960s, "modern experimental work on the production of models of autoimmune disease in laboratory animals is almost wholly the creation of Freund's complete adjuvant."[22]

At the beginning of the Cold War, Kabat and his colleagues in New York continued to use Freund's adjuvant to strengthen the immunological reaction of their monkeys to injected brain tissue, repeatedly inducing an allergic encephalomyelitis analogous to multiple sclerosis in humans.[23] But a rival scientist reported Kabat's past communist sympathies to the Federal Bureau of Investigation, leading to cancellation of his passport and withdrawal of his Public Health Service grants. As a result, in 1953 Kabat "had to kill off the only monkey colony in the world then being devoted to the multiple sclerosis problem."[24] He decided to concentrate henceforth more narrowly and carefully on the biochemical analysis of immunological processes.

Other scientists along the east coast of the United States quickly realized that, with Freund's adjuvant to boost the production of allergic encephalitis, experimental animals could readily be made models for human multiple sclerosis. In Boston, the neuropathologist L. Raymond Morrison, himself rapidly succumbing to pulmonary tuberculosis, managed to induce demyelination in guinea pig brains using injections of neural tissue with Freund's adjuvant. Morrison's brief career exemplifies the intense and tangled connections between neuroscientists and immunologists at Harvard and Columbia in the middle of the twentieth century. He had trained with Ferraro in New York, before moving to Boston where he worked with Putnam on multiple sclerosis.[25] Just before he died, he was collaborating with Freund on research into allergic encephalitis. An ambitious young assistant in Morrison's laboratory, Byron H. Waksman, eagerly took over the investigations. The son of the discoverer of the antibiotic streptomycin, Waksman had studied the chemistry of the immune response with Heidelberger at Columbia, learning the new techniques of electrophoresis and ultracentrifugation in preparation for his work on allergic encephalomyelitis in Boston.[26] Waksman became convinced that ordinary antibodies did not bear primary responsibility for the immunological reaction against normal brain tissue in allergic encephalomyelitis; rather, he believed its cause was delayed hypersensitivity, mediated by cellular components of the immune system, similar to the process giving rise to contact allergy and rejection of tissue grafts. In effect, it was like the tuberculin reaction, in which the skin becomes inflamed around an injection of mycobacteria, the extent of the lesion giving an indication of prior exposure to the tuberculosis germs. The response to tuberculin is delayed, and it was known by the late 1940s that cells, not antibodies, were involved, though their precise character and role had not yet been determined. According to Waksman, allergic encephalitis basically is a tuberculin reaction in the brain, though the eliciting antigens are normal brain constituents, not tubercle bacilli.[27]

At the start of the 1950s, many investigators realized that allergic encephalomyelitis and multiple sclerosis were autoimmune diseases, and it was becoming clearer to them what immunological reactions might be entailed. In reading the scientific reports of the time, it is easy to get caught up in dazzling stories of animal experiment, intricate technique, new adjuvants and reagents. Pieces of brain and other tissues are injected into a menagerie of docile animals; strange diseases appear; novel entities are created or discovered; concepts of self and other change shape. These romantic scientific quests can sweep up the avid reader, who may for a moment forget their real human referents. But human suffering continued to intrude on the scientific literature in the form of the clinical case report, like that in which a physician described a college student in her early twenties who consulted him when suddenly her right eye turned in and she developed double vision and dizziness. After a remission, she came back the following year when double vision recurred; subsequently she suffered episodes of numbness in her hands and face, and an ache on her left side prevented her studying. Within four years, "blurring of vision with almost complete loss of sight in one eye had been added to the clinical picture." Another case report described an insurance adjuster in his early thirties, who complained of numbness in his right leg and a burning sensation at the tips of his fingers. His walking had become unsteady and he showed a tendency to fall to the right and general lack of coordination. Within a year he was seeing double, with blurring of sight in the left eye, and was worrying about how his family would cope if complete disability overtook him. Then there was the teenager who noticed in his last year of high school that his right foot was always numb and sometimes his hands felt weak. Soon his right buttock also became numb, making sitting uncomfortable. His mother had to knot his tie because he could not coordinate his fingers.[28] What could it have meant to these and other sufferers of multiple sclerosis to learn that they were succumbing to an auto-allergy? What sort of answer was that to the questions troubling them?

❋ ❋ ❋

"Lupus is one of those things in the rheumatic department," Flannery O'Connor told the poet Robert Lowell in 1953; "it comes and goes, when it comes I retire and when it goes, I venture forth." During the previous four years, from her early twenties, the writer had endured recurrent fevers, muscle stiffness, joint pain, and a striking rash when her skin was exposed to the sun. "I have enough energy to write with," O'Connor continued, "and as that is all I have any business doing anyhow, I can with one eye squinted take it all as a blessing. What you have to measure out, you come to observe more closely, or so I tell myself."[29] Sheltering at the family farm in Georgia, O'Connor continued to write bleak misanthropic novels and short stories, but her disseminated lupus erythematosus grew worse. "I can work in the mornings," she wrote in 1957, but in the afternoons I can't do nothing but look at peachickens."[30] At her mother's insistence, she persevered with a pilgrimage to Lourdes, but it did no good. Aware that she was dying, O'Conner tried desperately to finish her short-story collection, *Everything that Rises Must Converge*. In 1964 she died, aged thirty-nine, from kidney failure, a complication of lupus.

Not long before O'Connor's diagnosis, hematologists Malcolm M. Hargraves and Robert Morton, with the technician Helen Richmond, reported in 1948 a peculiar cell, which they called the LE cell, in the bone marrow of the lupus patients referred to them at the Mayo Clinic in Rochester, Minnesota. For two years or more they had observed white blood cells in the process of engulfing and digesting some material, probably free nuclear matter left over from the breakdown of other cells. But the meaning of this phagocytosis, or cellular consumption of debris, baffled them, and their duties at the Mayo Clinic left little time for laboratory experiment.[31] Nonetheless, the presence of LE cells would become a diagnostic criterion of disseminated lupus—soon renamed systemic lupus erythematosus (SLE)—even though the pathological significance of these debris-laden cells remained obscure. Something had destroyed substances in the cell nuclei of the tissues of those suffering from lupus.

Initially the culprit was elusive, but within a few years scientists determined that the blood factor generating LE cells consisted of circulating gamma globulin—antibody. In the acute stages of lupus, the gamma globulin fraction of the blood proteins stimulated the production of LE cells; and once the gamma globulin fraction fell during clinical remissions, the LE cells disappeared.[32] The relationship with gamma globulin suggested an immunological basis for the disease, a supposition that further research soon confirmed. In a probative study, Swiss researchers mixed serum from lupus patients with a preparation of crushed cell nuclei. The nuclear material used up or absorbed all of the formative blood factor, or specific gamma globulin, thereby leaving the serum unable to generate any LE cells.[33] The results implied that the stimulating blood factor had acted like an antibody, binding to nuclear material and becoming utterly consumed or exhausted in the process.

Scientists at the Rockefeller Institute went further. Henry G. Kunkel and Halstead R. Holman confirmed that the provocative serum factor adhered to cell nuclei, but they also managed to elute the factor—in effect, to separate factor from nucleoprotein—and demonstrate it was an antibody. Once separated and exposed to another preparation of nucleoprotein, which was the presumed antigen, the antibody became completely absorbed, or used up. And when Kunkel and Holman fed the antibody-coated nucleoprotein to phagocytes, they quickly ingested the compound and formed LE cells. Toward the end of the experiment, the Rockefeller investigators realized it presented them with an opportunity to use a new fluorescent antibody detection technique, which traced the movements and activities of biological molecules, in this case gamma globulin. In the late 1940s, the Harvard immunologist Albert H. Coons had devised a means of labeling antibodies with fluorescent dyes: when the antibodies bound to their target molecules, serving as antigens, the tags brilliantly displayed the location of these substances under a fluorescence microscope. The fluorescent antibody technique enabled researchers to track the fate of antigens in tissues—and the activities of antibodies, since they could be treated as antigens, too, thus tagging one antibody to catch another antibody.[34] Accordingly, Kunkel and Holman prepared a fluorescent rabbit antiserum against normal human antibodies and then mixed it with LE cells: the nuclear material in the LE cells glowed revealingly while nuclear material in other cells stayed dull. Thus, they brilliantly illustrated that "the L.E. factor could be an autoantibody to nucleoprotein or deoxyribonucleic acid (DNA)."[35] Later identification and specification of these antinuclear antibodies offered further confirmation that SLE constituted another autoimmune disease.[36] "What we were witnessing," Holman recalled, "was evidence that an antibody was reacting with a nuclear antigen . . . and so we had an antibody to a nucleic acid arising in the disease."[37]

Reflecting on this discovery, Holman observed that "technological developments during the war created all sorts of new opportunities . . . and money became available to develop bioscience as it applies to clinical medicine."[38] The Coons immunofluorecence test was just one part of the new repertoire. Kunkel had recently purchased an ultracentrifuge, which allowed the sedimentation of serum proteins by molecular weight. Additionally, he devised starch block electrophoresis, enabling separation of proteins by electrical charge into a block, which then was cut into strips from which the different proteins could be eluted.[39] Kunkel was "very thoughtful about the ways you could make something work," Holman said. "He didn't spend much time on large, overarching concepts; it was how to make this work." When the results came through, Kunkel was so impressed that he sent Holman across town to let Kabat know they had discovered an autoimmune disease. Kabat, however, did not appear surprised.[40]

There is no evidence that O'Connor came to understand she suffered from an autoimmune disease, but a few others with lupus did begin in the 1950s and 1960s to engage with the idea that their immune system was assaulting their bodies. Henrietta Aladjem, a thirty-six-year-old Bulgarian-American housewife living in Boston, fell ill with fevers, headaches, fleeting rashes, and joint pains when visiting the Netherlands in 1953. Back home, she remained unwell, with fatigue, muscle stiffness, and waning strength. Her physician tried repeatedly to find LE cells, but the diagnostic test initially gave ambiguous results. By the late 1950s, though, Aladjem clearly was suffering from SLE, which, she learned, was a strange connective tissue disorder. Eventually, she heard about "antibodies that could attack a person's own body," and "some enzyme [that] causes damage to the DNA."[41] It was weird and surprising, but she wanted to know more. Around 1966, after reading the posthumous book, *Everything that Rises Must Converge*, Aladjem recalled, "My thoughts drifted to Flannery O'Connor, a writer I much admired." O'Connor had endured the same disease. "I could see her swollen face and hands; I could feel her pain and helplessness." Unlike O'Connor, Aladjem now knew her body "was forming antibodies to its own tissue," and she reflected, "What a horrible thought."[42]

❋ ❋ ❋

Flannery O'Connor was aware that her lupus derived from the general category of "rheumatic" disorders, classed with rheumatoid arthritis and other "connective tissue," or "collagen," diseases. The idea of collagen diseases proved remarkably attractive after the war, serving, it turned out, as a short-lived placeholder for concepts of autoimmune pathogenesis. In the 1940s, Paul Klemperer, a rheumatologist at Mount Sinai Hospital, New York, began to describe disseminated lupus, rheumatic fever, and rheumatoid arthritis as "diffuse collagen diseases," arguing that they displayed common alterations of connective tissue, especially abnormalities of the noncellular, or collagen, components.[43] The "fibrinoid alteration of the collagenous tissue" seemed to make some persons unusually sensitive to foreign proteins, thereby causing tissue damage. Klemperer and his colleagues concentrated on anatomical and chemical investigation, seeking to understand the character of the tissue defects, and left aside the problem of sensitivity. Collagen disease, Klemperer conceded, was "not a term applicable to diagnosis and certainly does not define the morbid process of the diseases grouped together."[44] In effect, this was a rather old-fashioned morphological and classificatory exercise, but its neatness and generality possessed broad appeal. Moreover, the idea of collagen disease eventually inclined other researchers, including immunologists, to assume that identifying the cause of one member of the class would have implications for allied conditions. If one so-called collagen disease, such as SLE, could be autoimmune, then why not another, such as rheumatoid arthritis?

Conducting routine Wasserman-type serological tests for syphilis in Oslo, Norway, in 1937, the pathologist Erik Waaler noticed a very strange result. As usual, Wasserman testing meant taking serum from a patient, adding a source of complement, and combining the mixture with the test antigen, a cardiolipin preparation, then adding sheep red blood cells with pre-bound anti–sheep blood cell antibodies as the indicator. If the patient's serum contained "syphilis antibody" then complement would be used up, or "fixed," in binding to the antigen-antibody complex; otherwise, complement would bind to the sheep blood cells, causing them to break apart, or undergo lysis, making the solution go red. But, with this particular serum, the sheep red blood cells agglutinated, clumping together. Waaler was intrigued. Something in the blood of the patient, an elderly man with rheumatoid arthritis, was inciting the sheep-cell agglutination. In order to investigate the phenomenon, Waaler took blood from other patients supposed to suffer from rheumatoid arthritis, and discovered that some, though not all, of these research subjects possessed the activating factor.[45] In 1939, he presented his results at a conference in New York, but the audience registered little interest. Returning to Norway, Waaler became caught up in the war, abandoning his studies of the mysterious rheumatoid factor.

Around 1948, Harry M. Rose, a bacteriologist at Columbia University, was investigating an outbreak of a febrile illness, probably caused by rickettsia, using a complement fixation test to confirm the germ's presence. His technician, Elizabeth Pearce, developed a fever and decided to test her own blood. Since her rheumatoid arthritis was troubling her, she set aside the tubes overnight, rather than checking them within an hour. The next day she found that the sensitized sheep red blood cells left with her serum had clumped together. When Rose and his colleagues tested the blood of patients with rheumatoid arthritis, they observed the same aggregation of the globulin-coated sheep red blood cells. Sadly, no one recalled Waaler's modest presentation some ten years earlier. Rose soon announced that he had discovered in the laboratory a means of diagnosing rheumatoid arthritis.[46] He had detected a distinguishing antibody in the blood of rheumatoid arthritis patients, a globulin (protein) that seemed to offer a clue to the disease's cause. Rose spent the remainder of his career trying to shrug off his Norwegian rival.

But what sort of antibody could this rheumatoid factor be? What was its pathological significance? Adept at centrifugation and electrophoresis, scientists in Kunkel's laboratory at the Rockefeller Institute in the 1950s determined that the rheumatoid factor was a high-molecular-weight globulin, an autoantibody binding to a segment of the gamma globulin molecule.[47] At first, it seemed almost preposterous that antibodies could be produced against other antibody components. "Our initial evidence that these 'factors' were antibodies," recalled Kunkel, "received considerable resistance even from immunologists. However, our simultaneous work on classical antibodies and their many different types placed us in a position of knowing more about human antibodies than anyone else, and acceptance of our proposals soon ensued."[48] Further studies demonstrated that these curious antigen-antibody complexes might become entrapped in the synovial membrane and cartilage of joints, causing irritation and chronic inflammation.[49] Like other "collagen diseases," rheumatoid arthritis therefore was an autoimmune condition. Excited by this discovery, Kunkel committed his laboratory to investigation of the role of circulating autoantibody-antigen complexes in a range of connective tissue disorders, including disseminated lupus. From the mid-1950s, his laboratory was busy converting the collagen diseases into autoimmune problems—even if Kunkel sedulously avoided the term.

Inspired in part by his Columbia colleagues Heidelberger and Kabat, Kunkel believed that protein chemistry would explain immunological processes. In order to elucidate complicated immunochemistry, he relied mostly on the ultracentrifuge and new methods of electrophoresis. Kunkel and his colleagues shared a "molecular vision of life," a conviction that basic physico-chemical principles and patterns operating at the molecular level could account adequately for life processes.[50] Their enthusiasm for protein chemistry persisted even as scientists interested in virus structure and the nature of heredity drifted during the 1950s into studies of nucleic acid.[51] While nascent molecular biology came to focus on the structure and function of nucleic acids, the recently discovered constituents of genes, immunochemistry would continue to identify and unravel the proteins making up antibodies.

Although committed to molecular analysis, Kunkel also prided himself on widening the passage between laboratory and clinic and maintaining one of the largest clinical units at the Rockefeller Hospital. He demanded "meticulous physical examinations, detailed hospital charts, and rigorously maintained prescribing principles," one assistant recalled.[52] On daily ward rounds, he often asked sharp questions about clinical presentations and therapeutic options. Kunkel believed the close connection of clinic and laboratory would dislodge conventional assumptions about rheumatic and other chronic diseases, stimulating new experimental approaches to these old medical problems. According to immunologist Baruj

Benacerraf, "he had the uncanny ability to use unique clinical material to make fundamental, broadly applicable discoveries."[53] Kunkel pushed his laboratory team hard, expecting researchers to work till early in the morning and on weekends and to show the same enthusiasm for immunological inquiry that he possessed. Like so many Cold War biomedical scientists, he sought to build a critical and competitive research culture in his laboratory. Yet, his involvement in medical care of patients meant that autoimmune research could never be reduced completely to molecular matters.[54]

⁂

Part of the postwar baby boom in biomedical science born around 1948 after a long gestation, autoimmunity was growing rapidly to maturity. In the 1950s, it acquired its name, which still sounded funny to some scientists and physicians. In 1963, the first textbook of autoimmune diseases could list scores of conditions with putative immunological origins, including multiple sclerosis, rheumatoid arthritis, disseminated lupus erythematosus and various other alleged collagen diseases, hemolytic anemia and other blood disorders, and many chronic inflammations of the thyroid, kidneys, gut, and liver.[55] Autoimmune contrivances, premised on an "allergic" sensibility, offered to many an attractive explanation for several puzzling tissue and organ pathologies. "One of the greatest developments in Medicine during recent years has been the growing recognition of the importance of processes in which the immune mechanisms of the body are, as it were, turned against the body's own components," wrote the textbook's authors, Ian R. Mackay and F. Macfarlane Burnet, from the Walter and Eliza Hall Institute in Melbourne, Australia.[56] During the previous fifteen years, extensive clinical and laboratory research, mostly conducted along the eastern seaboard of the United States, seemed to have established that diseases such as multiple sclerosis, lupus, and rheumatoid arthritis constituted autoimmune disorders, conditions in which the body's immune system ravaged its own tissues. Fundamentally, Mackay and Burnet wrote, autoimmune disease represented the failure to differentiate between self and non-self. And yet, despite the accumulated evidence, many clinicians and scientists remained skeptical, inclined to scoff at the idea that the body's regulatory processes might incite disease, that the immune system might not recognize its body as self.

Hemolytic anemia, in addition to multiple sclerosis and the collagen diseases, provided one of the more compelling illustrations of autoimmunity during this delicate stage in its conceptual development. For decades, physicians had been aware that the life span of red blood cells was limited, but in some cases their destruction, or hemolysis, could be accelerated, outpacing efforts by the body to replace them, leading to anemia. After the studies of paroxysmal cold hemaglobinuria conducted by Julius Donath and Karl Landsteiner early in the century, a few hematologists were prepared to believe that the body might be capable of producing some substance that attacked red blood cells, a hemolysin, though conventional opinion still attributed the anemia to an inherent flaw in the cells. Late in the 1930s in Boston, the physician William Dameshek wondered whether three of his patients could be producing hemolytic antibodies. To investigate this possibility, he and Steven O. Schwartz managed to get rabbits to make antibodies against guinea pig red blood cells; then they injected the serum into guinea pigs, causing hemolytic anemia. Their experiment implied that an immunological process was responsible for at least some cases of hemolytic anemia, but they failed to detect the actual antibodies against red cells.[57] They lamented the lack of a sensitive test of such immune activities. A few years later, Robin R. A. Coombs, a pathologist at the University of Cambridge, devised a means of amplifying, and thereby revealing, the effect of weak or "incomplete" antibodies by adding a rabbit anti-human-globulin serum. This ensured that red cells already coated or "sensitized" with weak antibodies would bond with the introduced antiglobulins, making them agglutinate. Using this method, Coombs and colleagues identified a means of detecting antibodies in maternal serum to the rhesus (or Rh) antigens of red blood cells, the cause of hemolytic disease in newborns.[58] When the Coombs test was applied to cases of suspected autoimmune hemolytic anemia, the red blood cells again clumped, indicating the pathogenic attachment of weak autoantibodies to the cell membranes. Accordingly, after 1951, physicians began to refer to *autoimmune* hemolytic anemia.[59]

The discovery of autoimmune thyroid disease also had far-reaching ramifications. After graduating from the University of Pennsylvania in the early 1950s, Noel R. Rose ventured north to work with the émigré immunologist Ernest Witebsky in Buffalo, New York. As Rose recalled, the "biological basis of the immune response was entirely obscure and intrigued me sufficiently to want to learn more about immunology."[60] Witebsky suggested that his acolyte study thyroglobulin, which seemed an excellent example of an organ-specific protein, limited to the thyroid gland, and presumably immunologically unique. The idea interested Rose. He prepared thyroglobulin from rabbit glands and then injected the material into other rabbits, but he observed no immune response. This did not surprise Witebsky, yet Rose persisted, wanting to know the reason why the animal could not produce antibodies to circulating thyroglobulin. He decided to inject rabbits with thyroglobulin from their own glands, having allocated them to three groups: one group with their thyroid glands removed so natural circulating thyrogloblulin could not override the effects of immunization, another with only part of the thyroid remaining, and the third with thyroid intact. Since the thyroglobulin antigen was scarce, Rose consulted with Freund in New York City, learning how to emulsify the thyroid extract in the complete adjuvant so that it should have greater immunogenic impact.[61] This time, the experimental results proved astonishing. All rabbits now made antibodies to their own thyroglobulin, and the remnant thyroid glands showed signs of severe inflammation, with abundant lymphocytes. "We were forced to conclude that we had actually produced an autoimmune response," Rose reflected, "and, even more exciting, an autoimmune disease, by this method of experimental immunization."[62]

Examining the thyroid glands of the autoimmunized rabbits, a collaborating surgeon thought the pathological picture closely resembled Hashimoto's thyroiditis, a human thyroid disease.

Rose attempted to get blood from Hashimoto patients, but it took two years for the local physicians to find twelve samples for him. Using a modification of the new Coombs test, Rose eventually determined that all twelve thyroiditis patients were producing antibodies against their own thyroglobulin.[63] Meanwhile, two scientists in London, Deborah Doniach and Ivan M. Roitt, heard about the rabbit research in Buffalo and set out independently to test the blood of chronic thyroiditis patients for autoantibodies. Doniach, a clinician, had already been looking at immunoglobulin patterns in thyroiditis, and Roitt, a biochemist, had been studying antibody production. They met at the Goodge Street tube station, around 1955. "We were both rather extrovert, I suppose," Roitt recalled, "and gabble, gabble, talking about what we were doing." Doniach speculated that the antibodies in Hashimoto's thyroiditis might be directed against a virus. "Jesus, I thought, or Allah or something," Roitt exclaimed, "it might even be [against] a thyroid component itself." Shortly, Doniach found some normal thyroid tissue and took blood from her patients. Roitt remembered it well: "I made an extract [of the tissue] and I set up these little tubes, stuck them in the plasticine, poured serum in and lo and behold, the good Lord gave us lines of precipitation!"[64] The patients were producing antibodies against thyroid tissue. Since it had proven easier to recruit people with Hashimoto's thyroiditis in London than in upstate New York, they published earlier than Rose.[65] "They had wonderful patients," Rose has said. "We had no patients of that sort and we had great difficulty finding them."[66] Years after these discoveries, Roitt insisted that if the scientists in Buffalo had "really thought that human disease might have been linked" to autoantibodies, "they should have turned heaven and earth over. . . . You network until you find somebody when you've got the possibility of something absolutely groundbreaking."[67] In any case, the positive findings of the British researchers added momentum to the search for other organ-specific autoimmune diseases. Soon they were referring more generally to the phenomenon of *autoimmunity*.[68]

The discovery of thyroid autoimmunity disturbed and challenged Witebsky, who insisted that Rose repeat the experiments and delay publication of his results. Once an associate of Hans Sachs, who had worked with Paul Ehrlich, Witebsky acquired in Germany a skeptical attitude toward claims of autoimmune processes, and he remained hesitant to admit that the immune system could destroy the tissues it had evolved to protect. "The field of autoimmunization and specifically its clinical application," Witebsky cautioned, "seems to have developed rather explosively and I have the feeling that a situation is developing which might be rather confusing and even damaging to the entire concept."[69] He persuaded his collaborators to include in their delayed article on Hashimoto's thyroiditis a list of criteria that should be fulfilled before attributing any disease to antibodies. These "postulates" stipulated the direct demonstration of circulating or cell-bound antibodies, recognition of the target antigen, production of antibodies against this antigen in experimental animals, and the development of pathological changes in the experimental animal that were the same as those in the corresponding human disease.[70] But the criteria failed as a containment strategy. Treated as ideals, Witebsky's postulates were invoked only rarely to exclude candidate autoimmune diseases. The pathological category continued to expand pragmatically in the 1950s and 1960s, encompassing more and more previously obscure medical conditions. As Witebsky reluctantly concluded, "the term 'immunopathology,' possibly considered by some as a contradiction in itself, . . . certainly looms as a new and wide field of investigative endeavor."[71]

❋ ❋ ❋

Although some scientists and clinicians shared Witebsky's reservations, other immunologists in the postwar years approached the search for autoantibodies with a "sense of unlimited possibilities," as J. W. N. Sullivan had once put it. "The floodgates were opened," Noel Rose recalled. "We all became autoimmunologists."[72] Enthusiastic scientists readily hurdled the old conceptual barrier of *horror autotoxicus*—the conviction that the body will not attack itself—entering a new territory occupied by hypersensitive, overreactive bodies whose normal immune responses could become pathological. Autoimmune mechanisms suddenly were conceivable and took on material form in immunology laboratories, explaining a variety of clinical conundrums. This investigatory pathway, formerly indistinct or blocked, soon was well marked—and over trodden, as Witebsky feared.

The scientists following this increasingly worn path often shared similar experiences, training, and aspirations. Many of them, including Waksman, Coons, and Kunkel, had served in medical roles during World War II, gaining confidence in their ability to command large numbers of patients and organize effective investigations and interventions, and learning the value of explicit instructions and protocols, the uses of standardization and improvisation, and a tolerance of drill and repetition. Most of them returned to elite academic medical centers along the eastern seaboard of the United States, where they directed laboratories and clinical research units, enhancing the articulation between scientific investigation and medical care. A number of scientists without medical training—some of them women—joined the research teams, thereby extending the conceptual and technical reach of the clinicians and marking a clear break with prewar convention.[73] Most early investigators of autoimmune mechanisms shared some connection with both established and emergent east coast medical and scientific networks, linking the institutional nodes of the Rockefeller Institute, Columbia-Presbyterian Hospital, and the medical schools at Harvard and the Johns Hopkins universities—along with the occasional outlier in Rochester, Minnesota, or Buffalo, New York. Often they had trained together; usually they attended the same conferences and wrote for the same journals; and all had read Landsteiner's *Specificity of Serological Reactions* (1933), especially its later English-language editions. Either themselves European émigrés, such as Freund and Witebsky, or trained at European centers, most of the postwar immunologists made sure to consort with older Continental scientific savants, such as René Dubos at the Rockefeller Institute or Louis Dienes at Harvard; a few before 1943 had even caught up with Landsteiner himself. Although the ideas that these researchers embraced were hardly new and unprecedented, the institutional matrix in which they found themselves and the career paths available to them differed significantly in scale and character from prewar conditions and opportunities.

Frequently, some of the major prewar disease burdens, especially rheumatic fever, poliomyelitis, and tuberculosis, prompted and gave urgency to their investigations, even though research trajectories diverged in the 1950s as other conditions proved more amenable to inquiries into immunological pathogenesis. Scientists in the large clinics of New York and Boston were gaining access to patients with less common ailments, such as the systemic collagen diseases, and rare chronic inflammations specific to certain organs. Sometimes they found that such conditions stimulated or ratified their interests in the pathological reactivity of the human body—in particular, the nondefensive activities of the immune system. The vast expansion of immunological tests and techniques, many of them obscure even to their practitioners, allowed clinician-scientists to propose unconventional autoimmune causes for many of these mysterious diseases. The new technical repertoire was extensive and diverse, as we have seen, encompassing Freund's adjuvant, LE cells, the Coons immunofluorescence test, the Waaler-Rose test, the Coombs test, electrophoresis, centrifugation, and complement-fixation tests, among others. These procedures permitted scientists to produce novel entities in the laboratory, unnatural amalgams that revealed natural functions and pathological processes, a sort of imaginative reconstruction, scaled down, of what was happening in human bodies.[74] Thus, human antibodies might combine with guinea pig globulin on sheep red blood cells to show, when scaled up, whether an autoimmune disease was occurring. From such laboratory chimeras developed a new grid of immunological intelligibility, a fresh way of understanding human disease causation, and, perhaps inevitably, a challenge to older concepts of what constitutes self.

Just as the allergic sensibility or thought style seemed ascendant, the word *allergy* began to lose its appeal for these researchers. *Allergy* implied some altered or extraordinary immunological reactivity, an abnormality or idiosyncrasy of immune function;

but increasingly the term was associated with excessive reactions to external agents—not to internal or self-antigens—and with the routine desensitization procedures of the allergy clinic.[75] Moreover, the kind of biological reasoning that allergy once implied gradually became less compelling for many immunologists, too. Studies of autoantibodies had been conducted on the terrain of allergy, dependent on a physiological understanding of the body; but by the 1960s, many investigators were retreating into biochemical analysis of causative mechanisms, seeking the molecular basis of immunological agency. The biological vision on which the field was predicated seemed about to dwindle into a more conventional ontological claim, one that posited an autoantibody specific to each putative autoimmune disease. The terminological drift from allergy to autoimmunity in part reflected this conceptual reconfiguration, representing a move away from speculation on biological variation toward more stable ontological ground. Therefore, no one was surprised when Rose and Witebsky discovered experimental *autoimmune* thyroiditis, and while linguistic inertia allowed experimental *allergic* encephalomyelitis to retain its historical associations until the 1970s, the empty signifier eventually was converted to the more realistic term *autoimmune*.[76] And yet, even as it was falling into the hands of immunochemists, the concept of autoimmunity also would become central to a more ambitious and encompassing biological understanding of immune function. That happened a long way from New York.

CHAPTER FOUR

The Science of Self

❋ ❋ ❋

Reflecting on his first visit to the United States, in 1944, the Australian microbiologist F. Macfarlane Burnet experienced a "new realization that an adequate scientific attack on almost any problem will provide a practical solution of it." Like many Australians during World War II, Burnet had awakened to the inevitability of a brash American future, though fond memories of more homely England lingered. As he returned to Melbourne in that year to become director of the Walter and Eliza Hall Institute, Burnet noted that in the United States "science is geared to the war effort," which prompted a "readiness to make large-scale human experiments when these seemed to be needed." Moreover, his travels had impressed upon him "the great value of team work by men with training and ability in widely different fields." In particular, the connections of laboratory and clinic in North America seemed unusually enterprising and productive. Burnet lamented that in Melbourne scientific research had "never been close enough to the actual clinical work and teaching of the hospital to provide the stimulus which one feels in the great teaching hospitals of England and America."[1] And yet, the Australian scientist, always analytical and reserved, viewed some American developments with disdain. In his trip diary, Burnet observed how competition for funding often made scientists "subservient," experimenters became dependent on elaborate equipment, and everywhere there was a "good deal of second order work."[2] All the same, Burnet would recall in his memoirs, "those three months in America represented the most intense and rewarding period in my life."[3] "The impression that America made on me was almost overwhelming."[4]

During the 1940s, Burnet was intent on discovering the nature of the immune response, which meant at the time learning how and why the body produced antibodies. Distrustful of simple chemical or structural models, the virologist sought a satisfying biological explanation of immune function. The conventional idea that an antigen, or foreign substance, served as a template for antibodies failed to excite him; it seemed biologically implausible. Rather, he came to wonder whether the body's capacity to discriminate between self and not-self constituted the real significance of immunity, or at least its priority. Burnet was as interested in discovering why we normally tolerate our own tissues as he was in explaining our ability to defend ourselves against pathogens. For him, recognition of self represented the fundamental biological problem. Through his struggles with the notion of the immunological self, Burnet would eventually secure the Nobel Prize in Physiology or Medicine (1960), establish his clonal selection theory as the major dogma of immunology, and elaborate an expedient conception of autoimmunity. Indeed, Burnet's speculations on the nature of the immune response brought autoimmunity from the margins of immunological research to its germinal center. For Burnet and his followers, autoimmunity implied an intolerant blunder and thus offered crucial insight into the normal means of distinguishing self and other.

On his first visit to the United States, Burnet was concerned mostly with delivering his lectures at the Harvard Medical School on disease ecology and catching up with the latest research on poliomyelitis and influenza viruses. But even then he made time to discuss variations of the "avidity" of antibodies with Michael Heidelberger and colleagues in New York.[5] On later visits, he talked more with Elvin Kabat, Isabel Morgan, and Harry Rose about the implications of their studies for autoimmunity.[6] Gradually shifting away from virus research, Burnet became increasingly interested in "opportunities for an approach to basic immunological problems."[7] Visiting in 1952, he again found himself "mildly infected with that mixture of tension and exhilaration . . . so characteristic of New York," where he was discoursing on immunological tolerance. A few years later, excited by new American and British work in immunology, Burnet was "seriously considering whether the time would be ripe to desert the field of influenza virus and move into that of antibody production."[8] On the verge of abandoning the virus studies on which his reputation rested, he planned to focus exclusively on the nature of immunological reactivity.[9]

❋ ❋ ❋

"The essence of allergic disease," Burnet wrote in 1948, "is its individuality." He regarded allergy as a vague and elastic category that could be summarized, perhaps too plainly, as "the inappropriate activity of immunological processes in the human body."[10] Allergy was a mistake of the immune system, he believed, often incited by foreign substances, sometimes complicating infective processes, and more rarely targeting the body's tissues, the body itself. According to those scientists he met in New York, there were "strong suggestions that the body is producing antibody against some of its own constituents," especially in the demyelinating diseases of the nervous system. This excessive reactivity and sensitivity of the immune response intrigued Burnet. "It is simply that in certain individuals the normal scavenging process, by which damaged cells are eliminated, is switched into an inappropriately immunological process—an antibody directed against some specific component of, say, damaged nerve cells is produced."[11] That is, the normal mechanisms of immunological tolerance had broken down. Such rare mistakes, with their disastrous consequences, confirmed his impression that "recognition of 'self' from 'not-self' is probably the basis of immunology."[12]

This insight had come gradually to Burnet during the previous decade. In the late 1930s, he was committed more modestly to explaining antibody production in terms of cellular function, thus countering prevailing biochemical or structural theories of antibody formation.[13] At the turn of the century, Paul Ehrlich had proposed the "side-chain" hypothesis, which suggested that cells discharged preformed antibodies, or side chains, specific to the antigen attaching to their surface membrane, and they could then continue to replicate the antibody indefinitely.[14] But Karl Landsteiner's discovery that the body mounted an immune response even to invented chemicals, substances never encountered naturally, seemed to imply—for those invested in rigid specificity—the need for an infinite repertoire of devoted globulins, most of them permanently superfluous. Hence, biochemists took the opportunity to insist on instructive models of antibody formation, whereby the antigen would act as a template for circulating globulin, which adjusted to fit the foreign chemical structure. In 1940, Linus Pauling at Caltech propounded the most compelling instructive method of antibody production, showing how complementary structures might rapidly generate replica antibodies, but Burnet remained skeptical.[15] He could think of no biological precedent for such a mechanism. Passing through Pasadena, California, in February 1944, Burnet tackled Pauling on this vexing issue, each of them "never quite understanding what the other was driving at."[16] To the Australian's irritation, most American scientists thought Pauling was right.

Burnet could not understand how popular instructive models accounted for some obdurate biological facts. Why is the antibody response amplified when the antigen is re-presented? How does the body remember the first antigenic encounter? What explained the continuing production of antibodies even when the antigen appeared to be absent? Why is the immune response so heterogeneous, with some antibodies demonstrating more avidity than others? Burnet believed that cellular activities must be involved in the process. In the late 1930s, he speculated that enzymes within unknown immunologically competent cells might "adapt" to antigens, which stimulated them to replicate, thus producing antibodies. These modified enzymes, which were proteins, would remain part of any descendant cells, providing an immunological memory of the antigen.[17] Burnet admitted the similarity to instructional models, but he could imagine no alternative. It seemed the immune response required training in order to recognize foreign materials and to respond appropriately. Once this learning process was complete, it could readily be reproduced through cellular means.[18] But Burnet had no idea of the mechanism of protein replication, even though this was widely assumed to happen. More disturbing, the adaptive enzyme theory was explicitly adaptationist and implicitly Lamarckian, postulating the inheritance of acquired characteristics, and therefore an affront to Burnet's Darwinist, or selectionist, sensibility. In other circumstances, Burnet preferred to argue that natural selection worked on genetic variation, but he tried to make the best of this exception, placing the requisite emphasis on training and adaptation.[19] He fancied that during embryonic development there occurred a "steady 'delegation of authority' from the nuclear genes which are necessarily constant in character in every cell, to cytoplasmic or at least non-chromosomal determinants."[20] Burnet tried to adjust to the fashion for environmentally modulated cytoplasmic inheritance, but it was an uncomfortable fit for him.

In these early attempts to reframe the immune response in terms of cell biology, Burnet resisted reference to tolerance, individuality, and self, yet the category of individuality at least was familiar to him. Since adolescence, Burnet had admired the popular tracts of Julian Huxley, whose writings had helped turn him into a biologist. The Australian acolyte followed closely Huxley's criticism of physico-chemical reductionism and his enthusiasm for eugenics, concern with education and training, and fascination with biological individuality. "The problem of individuality must be considered biologically," Huxley asserted in 1926, just when Burnet was coming alive to scientific research. "The concept of the individual is not absolute," the English biologist continued. "It is, in the first place, purely relative."[21] While Huxley had sensitized Burnet to the problem of biological individuality, exposure to the philosophy of Alfred North Whitehead in the 1940s would amplify his interest in the constitution of the self.[22]

In 1944 at Harvard, Burnet dined with Whitehead, whom he "revered as the greatest living philosopher."[23] Although conversation between the two taciturn men was strained, the scientist already was aware of Whitehead's arcane writings on the processual and embodied self. For the philosopher, the self emerged from the activity of encountering an objective world; individuality was the experience of the environment. According to Whitehead, the self was always in flux or becoming; the nature of the self, the actual entity, was to prehend or feel, to appropriate data from its environment to build itself up. *Process and Reality* (1929) is full of "self-experience" and "self-identity," "self-creation" and "self-formation."[24] It is likely that the philosopher was the source for another of Burnet's favorite words, *pattern*, which he used frequently beginning in the mid-1940s. For Whitehead, the self was a "complex unity . . . restricted under the limitation of its pattern of aspects." "What endures," he wrote, "is identity of pattern, self-inherited."[25]

While Burnet must have struggled with the philosopher's rebarbative language, he found nearby a helpful interpreter. In 1943, the immunologist acquired *A Contribution to the Theory of the Living Organism*, the strangely philosophical monograph of his cytologist colleague Wilfred E. Agar, also enthralled with Whitehead. In this biological pastiche of the philosopher's theories, Agar, professor of zoology at the University of Melbourne, described the self as "a unity of activity or process; and the constituents, or elements, of a process are events. . . . The unity of the self is also a unity of temporally successive events. The character of the self at the moment is the outcome of the whole of its past experiences. The activities, or processes, of the living organism constitute, therefore, a nexus of events." Agar emphasized the contributions of cell biology, embryonic development, and evolutionary processes, or events, to the creation of self. His book was "based on the premise that cells are subjects, and that the animal as a whole is a subject."[26] If nothing else, Whitehead's influence seems to have confirmed in the 1940s the importance of the idea of human biological individuality and given Burnet a distinctive term with which to express it: *self*.

And so, even as Burnet struggled to understand the immune response, he would keep returning to the problem of the biological basis of self. Slowly it dawned on him that the issue of self offered the means to assimilate immunology into general vertebrate biology. Toward the end of the 1940s, with his junior colleague Frank Fenner, Burnet began revising his adaptive-enzyme theory of antibody formation, this time focusing on immunological tolerance, the ability to discriminate between self and other. Burnet and Fenner incorporated into their new schema the recent finding of English zoologist Peter B. Medawar that an immune response caused the rejection of tissues transplanted from an unrelated individual. While it was widely known that donor skin grafts normally separated and died within a few weeks, the reasons for their failure had been contested. Medawar revealed an inflammatory process under the grafts.[27] Yet, the body's own skin obviously did not elicit this response. To Burnet and Fenner this implied that "almost every significant macro-molecule in the body may have an individuality that marks it off from similar macro-molecules in other individuals." All body cells must boast "self-markers," forestalling any immunological attack, mollifying the cells producing the antibodies.[28] But the central question remained: How might such tolerance, or absence of self-antigenicity, have developed?

Burnet and Fenner were convinced that recognition of self-markers must be imprinted during embryonic life. Previously, Burnet had concentrated on how antibodies are instructed to recognize foreign antigens, now he emphasized how the immune system comes to identify self. The studies of Wisconsin zoologist Ray D. Owen particularly impressed the Australian scientists. According to Owen, even nonidentical cattle twins always shared blood groups, since the uterus of cows permits connections between placental blood vessels, creating, in effect, a common placenta. That is, their blood mixed during embryonic life and no transfusion reaction, or immune response, occurred. Each twin tolerated the blood group of its sibling.[29] Burnet and Fenner believed therefore that something was happening in the uterus that ensured recognition of self, something that marked individuality. "If in embryonic life," they predicted, "expendable cells from a genetically distinct race are implanted and established, no antibody response should develop against the foreign cells when the animal takes on independent existence."[30] But, as Burnet later admitted, "it is an untidy theory, because it postulates happenings for which there is only the slightest evidence."[31] The mechanism of immunological tolerance was still conjectural.

❋ ❋ ❋

Born in rural Victoria during the South African Boer War, before federation of the Australian colonies, Burnet belonged to an earlier generation than most of the Americans investigating autoimmunity. A social isolate, he always found solace in the study of natural history, even as he made his career in medical research.[32] Before World War II, he had conducted important studies of bacteriophage, poliomyelitis, influenza, and other viruses, developing an international reputation for his acute biological insight and ecological orientation, though he was based mostly at the Hall Institute in Melbourne, distant from any other major scientific center.[33] When Burnet was appointed director of the institute in 1944, it was moving to a new laboratory building adjacent to the Royal Melbourne Hospital, for which it performed pathological services. During the war, medical staff from Case-Western Reserve Medical School in Cleveland, Ohio, took over the hospital, turning it into a United States military operation. Burnet regarded this occupation hopefully, as signaling the more general shift of Australia away from Britain and toward America. In the postwar years, the Hall Institute retained its small scale, with fewer than thirty research staff, most of them investigating viruses, with meager funding and resources, and a prevailing sense of detachment—both from scientists in the Northern Hemisphere and from the sprawling, respectable, dull metropolis surrounding it. This seems to have suited Burnet. "I preferred to keep the Institute strong on ideas and weak on equipment," he recalled—so long as they were his ideas.[34]

In the 1950s, Burnet's dedication to virus research continued to dwindle as the field became more molecular in style and increasingly dependent on tissue culture techniques. Drawn to speculative immunology, he tinkered some more with his adaptive-enzyme explanation of antibody formation, reframing it this time in terms of information theory. To understand the immune response, he wrote in 1954, one needed to know about "replicating patterns which carry information or instructions from one part of a cell or organism to another." "It is in the spirit of the times to believe that we shall soon see the conscious development of a 'communications theory' of the living organism."[35] He was searching in vain for a "language" that would "function in these twilight zones where chemical description of molecular structure and function breaks down." He yearned for a "communications theory of the cell" to counter the "futility of the physico-chemical approach."[36] He claimed, "It is reasonable to think of the biological function of the [antibody] protein as analogous to the meaning of a paragraph." Making sense of a paragraph "can be regarded as a biological process analogous to the 'recognition' of a functional protein based on some form of complementary pattern."[37] He realized that his reference to "meaning" indicated how far he was straying from strict information theory; he feared that the living organism contained too much information for useful analysis; and he wondered more generally whether his immunological thinking had reached an impasse. "At the moment the Burnet-Fenner theory of antibody production is at its top," he confided to his wife in 1956, "but I am already beginning to see it has got to be replaced by a better one quite soon."[38]

Information theory continued to appeal to Burnet, even if it did not mint out a new theory of the immune response, as he had hoped.[39] All the same, he would remain fond of notions of control, communication, feedback, coding, message, memory, replication, and pattern—the last an old favorite. In particular, the ideas of émigré Viennese radiologist Henry Quastler proved stimulating. Quastler persuaded Burnet that information was

> related to such diverse activities as arranging, differentiating, messaging, ordering, organizing, planning, restricting, selecting, specializing, specifying, and systematizing; it can be used in

> connection with all operations which aim at decreasing such quantities as disorder, entropy, generality, ignorance, indistinctness, noise, randomness, uncertainty, variability, and at increasing the amount or degree of certainty, design, differentiation, distinctiveness, individualization, information, lawfulness, orderliness, particularity, regularity, specificity, uniqueness.[40]

The cybernetic approach also reinforced Burnet's belief in the dynamic figure of the self. As Norbert Wiener wrote in 1950, "to describe an organism, we do not try to specify each molecule in it, and catalogue it bit by bit, but rather to answer certain questions about it which reveal its pattern." It could have been Burnet's credo. "The biological individuality of an organism," Wiener went on, "seems to lie in a certain continuity of process, and in the memory by the organism of the effects of its past development."[41] Burnet took this as confirmation that the immunological self must share the nature of communication, not identity of matter.

Burnet's continuing attachment to information theory and cybernetics would give him what historian Donna Haraway calls more generally an "operations research approach to biology."[42] As a biological theorist, Burnet drew on concepts and metaphors circulating during the Cold War. He translated an older militaristic discourse on defense and attack into the new language of control, communication, recognition, tolerance, and surveillance. He emphasized relational subjectivity and worried about rigid conformity to external models. His immunological manifestos expressed concern about the porous boundaries of self and other, and fears of overreaction and hypersensitivity. Finely attuned to the American and Australian politics of the Cold War, Burnet intuitively proposed matching metaphors and concepts—figures and tropes that derived from the prevailing thought style.[43]

❋ ❋ ❋

At a certain point in the investigation, Peter Medawar regretted his boast that he could distinguish fraternal and identical cattle twins by observing how long reciprocal skin grafts lasted. At a farm outside Birmingham, England, he and Rupert E. "Bill" Billingham discovered, to their surprise, that all cattle twins accepted skin grafts from each other, even animals of different sexes, which could not possibly be genetically identical. None of them rejected the grafts. The zoologists "could not make head nor tail of our results."[44] But then, "browsing in an exciting-looking book"—Burnet and Fenner's *Production of Antibodies* (1949)—they found the answer.[45] As they read about Owen's studies of cattle twins, they realized that, since the animals were transfused with each other's blood before birth, they developed an immune tolerance to each other's tissues. Recovering from his embarrassment, Medawar claimed his research had confirmed Burnet's self-marker hypothesis.[46]

"Thank God we've left those cows behind!" Medawar said when he moved soon after to University College, London, with Billingham and a sharp graduate student, Leslie Brent.[47] But their bovine experience had made them confident they could now induce in laboratory animals the same tolerance that the unique placental structure produced in cattle. They tried inoculating mouse embryos with foreign cells and discovered that the adult mice had acquired immune tolerance to the inoculum.[48] Since the 1880s, medical scientists had sought the means to generate immunity to disease; but now the British zoologists were describing how immunological *tolerance* might be induced. In effect, they had worked out how to break down the natural barrier preventing the transplanting of genetically foreign tissues. As one of their colleagues remarked, "it gave new hope to those who were trying to make human organ transplantation possible."[49]

Burnet, of course, was pleased that Medawar and colleagues could demonstrate that immunological tolerance developed in embryonic life, even if other aspects of his theory continued to trouble him. He found the ebullient, spirited Medawar sympathetic to his own biological inclinations; they both wanted to understand how self could be distinguished from not-self; and they shared a distaste for physico-chemical reductionism and complicated equipment. Yet others in Medawar's group evinced little rapport with Burnet. Brent recalled that he "used to think of Burnet as 'a dry old stick,' certainly not given to small talk, with a less than glowing personality." The reticent Australian scientist "had a genius for picking up experimental studies that suited his ideas and ignoring those that opposed them."[50]

❋ ❋ ❋

Meanwhile, in Copenhagen, Niels K. Jerne was contemplating variations in the avidity of antibodies—variations, that is, in the engagement of antibody and antigen—and the challenge they presented to efforts at standardization. Haughty and snappish, Jerne chafed against the tedium of routine serological work in his Danish laboratory. He imagined himself too urbane and bohemian for ordinary experimental labor, as a figure like Søren Kierkegaard, mistakenly fallen among scientists.[51] But changes over time in the binding of antitoxin to diphtheria toxin did excite Jerne's philosophical interest. The alteration in avidity suggested to him that a heterogeneous population of antibodies was responding to the antigen. It also meant that antibodies could differ in the strength and character of their bonds to the antigen, that they did not have to match it precisely to display affinity. Such variation in antibody sensitivity and reactivity implied that the body would not require, after all, an infinite repertoire of antibodies to respond specifically to every possible foreign substance. Jerne believed that it was biologically feasible for an organism to have enough cross-reactive preformed antibodies to deal with any threat. Rigid instructive theories of antibody formation now seemed spurious.

One day in 1954, as he strolled across the Knippel Bridge in Copenhagen, Jerne suddenly worked it all out.[52] According to his "natural-selection theory" of antibody production, the foreign antigen binds to a preformed, natural, antibody with a complementary configuration, as it circulates in the bloodstream, and then the complex enters cells that replicate and multiply the antibody. "The role of the antigen is neither that of a template nor that of an enzyme modifier," Jerne wrote. "The antigen is solely a selective carrier of spontaneously circulating antibody to a system of cells which can reproduce this antibody."[53] An admirer of the British statistician and neo-Darwinian R. A. Fisher, Jerne was especially pleased with the notion that the antigen exerted a positive selection pressure on the immune system. It felt more biologically sound than the prevailing instruction hypothesis. This selective hypothesis echoed faintly Ehrlich's side-chain theory, though Jerne omitted any reference to his predecessor. Jerne also was evasive about the actual mechanism of antibody production. "Somewhere in the beginning," he wrote, "we have to postulate a spontaneous production of globulin molecules of a great variety of random specificities in order to start the process."[54] He could only speculate on how, when, and where the repertoire of random antibodies formed. Nor was it clear why the delivery of an antibody to the right cells should prompt them to synthesize further proteins. While Jerne's theory conformed well to many biological aspects of the immune response, its genetic and chemical basis seemed shaky.

Burnet, after speaking to Jerne early in 1957, recorded, "I did not find myself at all sympathetic to his approach, mainly because I could not see that this approach could cover the self-not-self differentiation crucial to all immunological theory."[55] But Burnet would not let it go. The natural selection hypothesis displayed a "verve and a sweep that made the self-marker ideas I was struggling with at the time look clumsy and artificial."[56] "I came back to Australia," he recalled, "pondering heavily on why Jerne's theory was so attractive, though obviously wrong." As a biologist, he particularly appreciated its Darwinian framework, though evidently it would need major adjustment and modification. For months he played with it. And then, "rather suddenly, 'the penny dropped.'"[57] Or, as Jerne later wrote, "I hit the nail, but Burnet hit the nail on its head."[58]

❋ ❋ ❋

As Burnet was contemplating Jerne's natural-selection theory, clinical investigations at the Hall Institute impressed upon him the need to account for pathologies of immunity in any theory of antibody formation. Inspired by the ramifying clinical research enterprise in the United States and aware of the institutional advantages in forging closer ties with the adjacent Royal Melbourne Hospital, Burnet had established the Clinical Research Unit (CRU) in 1946. During its first decade, clinician-scientists concentrated on diseases of the liver and gastrointestinal tract, developing a new technique of stomach biopsy and defining the lesions of chronic gastritis.[59] It was creditable research and clinically serviceable, but hardly calculated to excite Burnet. That would change in 1955.

Some CRU scientists became interested in a disease then called chronic active hepatitis, a recently recognized inflammatory and destructive disease of the liver, mainly affecting young women. "The course was progressively downhill," one physician wrote; "jaundice became a permanent feature . . . bleeding episodes became more frequent . . . . Alternatively these patients became permanently bedridden owing to their dropsical enfeebled state, and finally lapsed into coma. . . . The time from presentation to death has varied between six months and two years."[60] At first everyone assumed it must be a chronic viral infection, but Burnet cautioned that viral infection was seldom chronic.[61] Another feature was that all the afflicted showed exceptionally high concentrations in their blood serum of gamma globulin, the antibody protein, which suggested some unusual immunological activity.[62] In 1955, CRU scientists described a woman with chronic active hepatitis whose liver biopsy contained dense infiltrates of lymphocytes and derivative plasma cells, as well as Malcolm Hargraves's lupus erythematosus (LE) cells in the blood—implying an association with, or a relation to, the suspected autoimmune disease. Accordingly, they wondered if this meant that "an abnormality of the antibody-producing mechanism" or modification of the "self markers" of liver cells might explain this "lupoid" hepatitis.[63] Evidently, Burnet was proffering advice.

As it happened, a visiting researcher at the Hall Institute, D. Carleton Gajdusek, was fiddling about in one of its laboratories, trying to devise a new procedure for detecting hepatitis virus. Prodigiously gifted yet temperamental and troubled, the young American admired Burnet's "deeply analytic approach to problems of virology and . . . astounding command of all biology and basic life science."[64] Therefore he had included Melbourne on his "medical snooping" itinerary, a global research tour—despite Burnet's obvious discomfort with the unwanted acolyte's flamboyance and unreliability.[65] Sensing the rebuff, Gajdusek had begun his own research project at the institute, crafting a technique to detect hepatitis virus. He obtained some liver tissue from a patient who had died from acute viral hepatitis to serve as the source of antigen, and he used a complement fixation reaction to test for antibodies against homogenized liver in the blood of patients suspected of harboring the infection. The test cases of acute infectious hepatitis proved only weakly positive. Then, Ian R. Mackay, a young physician-scientist recently appointed to the CRU after a stint in Seattle, suggested using as controls the serum from patients suffering from other liver diseases, which included chronic active hepatitis.[66] This time the results were dramatically positive, not only using the virally infected liver as antigen, but with normal liver tissue, too. It looked as though this group of patients was overproducing antibody against normal tissue, not against a virus—Gajdusek and Mackay had, in effect, identified chronic active hepatitis as an autoimmune disease.[67] "I believe that my discovery will open up a fully new approach and a completely new field of human pathological physiology," Gajdusek wrote excitedly to an American friend.[68] "It is the first really original and important thing I have turned up," he told Joe Smadel, his mentor, who was then at the National Institutes of Health in Bethesda, Maryland.[69]

Burnet was fascinated. "If we were to understand such things," he recalled, "it was clear that any adequate interpretation of normal antibody production must have something to say, also, about the pathology of immunity."[70] For a theory to be "accepted as the basis on which to build an intellectual interpretation of classical immunology it must also be the basis of understanding immune processes when they go wrong."[71] Any satisfactory theory of immune function would need to explain not only tolerance but also autoimmunity—its failure. Burnet did not see how Jerne's natural-selection hypothesis could do so without modification.

❋ ❋ ❋

"It gradually dawned on me," Burnet wrote, "that Jerne's selection theory would make real sense if cells produced a characteristic pattern of globulin [antibody] for genetic reasons and were stimulated to proliferate by contact with the corresponding antigenic determinant."[72] Jerne had proposed that circulating antibodies bound to antigens, then stimulated cells to produce more antibodies, but Burnet thought it unlikely that molecules of partly denatured antibody could make a cell do anything. Rather, having mused on the matter throughout 1957, he decided to displace the selection pressure onto immunologically competent cells—assumed to be lymphocytes—and not onto antibody. Each lymphocyte must be genetically programmed to produce a specific antibody, Burnet figured. When receptors on its surface recognized the right foreign pattern, the lymphocyte would start proliferating, forming a clone of cells producing the antibody. Cell biology explained immunological memory, which caused the more potent secondary reaction to an antigen and permitted continuing production of antibodies even in the absence of provocation. The "clonal selection theory" was "basically an attempt to apply the concept of population genetics" to immunologically competent cells. It refigured the cells of the immune system as a Darwinian microcosm.[73] To Burnet's relief, clonal selection discarded any instructive or Lamarckian elements of earlier theories, including his and Fenner's adaptive enzyme hypothesis. "At no time does information enter the cell from outside," Burnet noted defensively. "The experimental change in the environment allows the emergence into activity of what was formerly only a latent capacity."[74]

For "ease of exposition," Burnet in 1957 chose to call the immunologically competent cells lymphocytes, though he continued to make occasional reference to more enigmatic mesenchymal and reticular cells. Leslie Brent recalled, "Lymphocytes certainly had been regarded as the 'Cinderella' of the immune system—generally neglected and an object of the most profound ignorance and even disdain."[75] But, also in 1957, James Gowans, working with Howard Florey at Oxford, discovered that small lymphocytes were circulating from the blood to the lymphatic system, including lymph nodes, and then back to the bloodstream, making them eligible antibody producers.[76] Moreover, Morten Simonsen determined the same year that white blood cells, mostly lymphocytes, in grafts could mount an immune response against host tissues.[77] When Burnet came to deliver the Abraham Flexner Lecture at Vanderbilt University, Tennessee, in 1958, he confidently declared there was no longer "any serious objection to the view that the main contribution of circulating antibody comes from cells with the histological character of immature plasma cells"—cells that almost certainly derived from lymphocytes.[78] The studies of Gowans and Simonsen had "made it admissible to postulate that a lymphocyte appropriately stimulated could give rise to a clone of dependent cells."[79] As Burnet put it, the clonal selection theory had "highly relevant implications for the general function of the lymphocyte."[80]

Burnet suggested that the body's "immunological pattern" was formed during embryonic life, through the process of somatic mutation in lymphocytes, which randomized the coding of the globulin molecules. The process of clonal selection after birth could "mould the population of mutants to provide the exquisitely adapted structure of immunological surveillance and defence."[81] For this mutant defensive population, the foreign antigen was an "expression of difference," exciting lymphocyte proliferation.[82] Burnet believed that clones carrying receptors for self-determinants, for the body's tissue antigens, normally were eliminated or inhibited at an early stage of life. Therefore self-recognition, or tolerance, was not hereditary; rather, he wrote, it "seems to develop as a secondary process sometime during embryonic life." "'Forbidden clones' that match 'self'-antigens," Burnet speculated, "would be eliminated as they arose," perhaps destroyed by embryonic contact with the antigen.[83] It appeared that pathological activation of a "forbidden" or inhibited clone—a breakdown of the tolerance mechanism—might explain autoimmune disease.

While Burnet was formulating his clonal selection theory early in 1957, he received David W. Talmage's article briefly mentioning a similar selective hypothesis. An academic allergist at the University of Chicago, Talmage proposed a cognate cellular selective process to modify Jerne's theory, though he avoided the issues of generation of antibody diversity and induction of tolerance.[84] Burnet acknowledged Talmage in the announcement of his clonal selection theory, but the affinity of their theories became a lifelong embarrassment. Talmage thought Burnet "truthfully had developed the idea before he received my paper." The Australian scientist "gave it a very catchy name, 'clonal selection.'"[85] "My only consolation," Talmage wryly remarked, "is the knowledge that Burnet's theory is a 'clone' of mine."[86] The American would continue to play the role of Alfred Russel Wallace to Burnet's alter ego, Charles Darwin.

For several years, Burnet observed, "nearly every attempt to disprove the theory was successful in doing so, but every new discovery in immunology . . . fell sweetly into place to strengthen and widen the approach."[87] A brilliant young researcher from the University of Wisconsin, Joshua Lederberg, visited the Hall Institute in the late 1950s and gave strong support to clonal selection. As the discovery in 1953 of the double-helical structure of deoxyribonucleic acid (DNA) had galvanized scientists, focusing their attention on molecular biology, Lederberg attempted to align Burnet's theory more closely to the recent discoveries in genetics. The American visitor admired Burnet's "uncanny biological vision" but felt his ideas needed to be "translated into DNA language."[88] Additionally, Lederberg observed that lymphocytes kept mutating as they proliferated in adult life, implying the possibility of inducing tolerance in early cell development, not only in the organism's development.[89] Lederberg, along with a "young, audacious postdoctoral fellow," Gustav Nossal, devised an experiment to test whether each lymphocyte produced just one kind of natural antibody, a prediction of the clonal selection theory.[90] After many "tedious micromanipulations," their results indicated that a single lymphocyte generally issued one type of antibody, though others disputed this finding.[91] Nonetheless, evidence was accumulating of the explanatory power of clonal selection.

Many years later, in an address on "the complete solution of immunology," Jerne generously declared that, "in principle, immunology was solved in 1957 when Burnet published the 'Clonal Selection Theory of Acquired Immunity.'"[92] But many immunologists remained skeptical. "It is very difficult for a chemist to accept this hypothesis," wrote Felix Haurowitz, "because it cannot be translated into the language of chemistry."[93] According to Gerald Edelman and Baruj Benacerraf, "instructive theories appear to have been elected by immunologists properly concerned with the chemical basis of specificity. Those concerned with the biological events of antibody synthesis favored the selective theories."[94] Gradually, the biologists came, for a period, to dominate discussion. And so, at the 1967 Cold Spring Harbor Symposium on Quantitative Biology, Burnet was able to savor his biological triumph.

❋ ❋ ❋

In 1961, Burnet detected "acute interest among physicians in those diseases which in whole or in part result from a misdirected immunologic attack on some of the body's own components." The "new immunology," he pointed out, was no longer concerned with defense but with "an altogether subtler process," discrimination between self and not-self. The failure to recognize self—autoimmunity—thus represented either a mutation in an immunologically competent cell or "the failure of a homeostatic mechanism," just like older ideas of fever. A forbidden or inhibited clone of lymphocytes, reactive to self, somehow escaped repression or control. Autoimmune disease, the dark side of immunity, is "protean" in character, "varying not only in intensity and in course but also in the organs or systems predominantly involved."[95] Burnet believed this "not an easy concept to explain either to laymen or physicians." The best analogy he could adduce was a "mutiny in the security forces of a country."[96] On another occasion, he likened "communication and control of the immune system" to the "control of crime or delinquency or the economics of an industrial society."[97] Of course, he always allowed, "one cannot discuss autoimmune disease without getting into deep water philosophically."[98] But Burnet was a confident swimmer.

Toward the end of his career, Burnet became fixed on laboratory studies of autoimmunity, which he regarded as the "most important unexploited field for medical research at the present time."[99] Around 1955, he had heard of some black mice, bred in the course of cancer research at the University of Otago, Dunedin,

New Zealand, that had spontaneously developed an autoimmune hemolytic anemia. Another strain, bred from the New Zealand Black (NZB) mice, gave positive LE cell readings and suffered kidney damage similar to the lesions of human lupus erythematosus.[100] The natural occurrence of these diseases in the animals, without the aid of adjuvants, added credibility to the autoimmune concept. From 1959, Burnet and Margaret Holmes assiduously investigated these animal models of autoimmune disease, yet the results proved disappointing, as they failed to illuminate the basis of autoimmunity.[101] Although an accomplished experimenter, Burnet often enjoyed greater success in theoretic endeavor.

❋ ❋ ❋

For a time, Burnet managed to impose his own broad biological framework upon immunology, displacing narrowly structural or chemical approaches. Obsessed with the distinction between self and other, he inadvertently made the nature of autoimmune disease crucial in understanding basic immunological function. Usually, such clinical problems were little more than prompts for his laboratory research, but the propensity to mount an immune response to the body's own tissues—the excessive or misguided reactivity of bodies, manifesting as disease—raised fundamental questions concerning methods of self-recognition and tolerance. In order to comprehend normal immunological function, he believed, one must explain its mistakes and failures. "It is a truism," he claimed, "that the best test of a physiological concept is its application to pathological conditions."[102] Autoimmunity thus became for Burnet and his followers a major problem in general biology, not simply a refractory clinical challenge.[103]

"It is a pity," Burnet had reflected in 1952, "that so much of modern medical research is so deeply concerned with the minutiae of chemical and physical functioning within the body that there are few men of the highest quality who have any time to look at the broader biological aspects of health and disease."[104] He liked to think of himself as a biological prophet—largely unheeded—emerging from the Australian desert. Increasingly pessimistic about reductionist tendencies in biomedical research, he railed against the molecular vision of life. Quixotically, he asserted in 1966 that molecular biology was "very largely a laboratory artefact that has never been brought into useful relation with biological realities."[105] At the same time, Burnet's major biological insight, the clonal selection theory, would depend on these molecular studies for its eventual validation. According to immunologist Melvin Cohn, "the acceptance of 'selectionism' was driven by advances in molecular biology."[106] Even Burnet's protégé Nossal admitted that "the greatest force for the acceptance of the theory was the progressive unraveling of the molecular side of the antibody story, both the protein chemistry and the molecular genetics."[107] Speaking at Cold Spring Harbor, New York, in 1967, Jerne described two prevailing "schools of thought" in immunology: the "cis-immunologists" consisted of biologists interested in the functional aspects and patterns of immunological responses to antigens, whereas the "trans-immunologists" focused on physical features of antibody structure and the chemical or structural basis of specificity. Sadly, Jerne observed, "a cis-immunologist will sometimes speak to a trans-immunologist; but the latter rarely answers."[108] By 1967, the trans-immunologists evidently were taking over again, to Burnet's chagrin. His work had incited merely a biological moment in theoretical immunology, and not an era. The rest would be chemistry—or so it seemed to the aging biologist.

CHAPTER FIVE

Doing Biographical Work

❋ ❋ ❋

What did it mean for a person to become "autoimmune" in the second half of the twentieth century? For physicians in the 1960s, the novel causal mechanism, while still enigmatic, could sharpen the diagnostic image of a patient's vague, indefinite illness; often it suggested new treatments, focused on suppressing the immune response; and sometimes it improved the prognosis of many of these debilitating conditions. Validated as a pathogenic pathway, autoimmunity brought rheumatic, neurologic, endocrine, and other patients—once scattered across the varied contours of internal medicine—into contact with the emerging specialty of clinical immunology. The new specialists opened a conduit between the medical routines of the hospital ward and the arcane procedures of the immunology laboratory. They proved adept at applying and interpreting the proliferating serological tests of immunological reactivity; over time, they came to know well the natural history of autoimmune diseases; and gradually they learned to monitor and adjust new medications, to watch for the many side effects, and to offer the necessary symptomatic relief when all else failed. From the 1960s, the diagnosis of an autoimmune disorder would set the sufferers of these diverse illnesses on a distinct trajectory within the modern medical system. The discovery that an autoimmune process was occurring within the body changed the sick person's illness "career": she, or more rarely he, recruited a new group of specialist physicians and carers and acquired an array of new tests and a regimen of new, often experimental, treatments.[1] Becoming autoimmune exerted an influence, sometimes determinate, often subtle, on the way a person might travel through illness.[2]

Still, autoimmune disorders, including multiple sclerosis, systemic lupus erythematosus, rheumatoid arthritis, and chronic hepatitis, continued to sit within older, time-worn disease categories, and it remained uncommon for patients to lay claim to an "autoimmune" disease or to define their medical problem as an "allergy" to the tissues and organs of their own body. They were experiencing chronic illness—debilitating, frustrating sickness that lasted year after year, even if occasionally remittent or wavering in its otherwise seemingly inexorable course. They took medications that might control the disease for a while, but often made them feel sick and ugly and vulnerable. "Acting like a sponge, illness soaks up personal and social significance from the world of the sick person," writes medical anthropologist Arthur Kleinman.[3] Few autoimmune patients were inclined to think much about their immunological dysfunction. While the practices of clinical immunology were transforming or taking over their lives, the concept of autoimmunity usually evaded, or perhaps repelled, their imaginations. They had other ideas. They struggled on, attempting to reassert control and find direction in an existence frequently dominated by long-term illness. "The body—its corporeal or physical aspect—was no longer an efficient and reliable instrument," sociologists reported after a study of multiple sclerosis patients. "It seemed to set its own agenda and have its own requirements, which competed with and inconvenienced preferred activities."[4] In suffering such an illness, one became estranged enough from one's body without delving into the nature of autoimmune pathology.

Immunology, as Macfarlane Burnet asserted, may have become the science of self, but its impact on the patient's sense of self was surprisingly indirect and elusive. To be sure, Henrietta Aladjem, a lupus sufferer and health activist in Washington, D.C., even in

1972 could express horror that her antibodies were destroying her own tissue, failing to recognize her self.[5] But such immunological speculation was extremely rare, likely to be found mostly in more medically involved patients. In general, sufferers from long-term illness experienced a visceral, mundane assault on their residual sense of self.[6] Over time, suffering eroded self-esteem, confidence in independent agency, hopes of achievement, and feelings of control over one's life. Those afflicted feared dependence and stigma and isolation. Daily, people with autoimmune diseases found themselves dealing with the uncertainties and irregularities of chronic illness. There were times when they imagined themselves immersed in illness. Life could seem over-determined by illness. The self might retreat into the illness. For many, chronic illness found expression in a language of loss—in particular, the loss of self—a language more meaningful and profound, if less scientifically fortified, than the discourse of self and not-self articulated by immunologists.

In this chapter, we touch on the experience of living with an autoimmune disease, especially in the 1960s and 1970s when the idea was novel, though our sources are scanty and the patients' voices almost inaudible. Necessarily limited and partial, this effort represents a conceptual history from the patient's view. We seek, in particular, to illuminate the dark passages of diagnosis and therapy, places where the experience of chronic illness intersects or becomes entangled with increasingly standardized biomedical trajectories. These intersections are the most common points of communication between patients and physicians, and sometimes provided opportunities for transformation or adaptation. Diagnosis and treatment can alter the trajectory of affliction and modify the experience of illness—thereby changing a life. Occasionally, the challenge of diagnosis and treatment might cause physicians to reflect more critically on conventional understandings of a disease. More rarely, intense and sympathetic engagement with patients has changed their lives, too.

❋ ❋ ❋

Modern medical diagnosis is predicated on disease specificity, the assumption that some singular disease entity is there to be found.[7]

But the protean manifestation of autoimmune processes often defies simple attribution of disease categories. When do diffuse aches and pains, a little morning stiffness, and a swollen finger coalesce into rheumatoid arthritis? At what point do intermittent tiredness, sore joints, and headache become systemic lupus erythematosus? What is the threshold at which muscle weakness, fleeting double vision, and loss of dexterity turn into multiple sclerosis? Occasionally a striking symptom or sign might announce the disease. Transient blindness in one eye strongly suggests a diagnosis of multiple sclerosis; deformity of the fingers, rheumatoid arthritis; a butterfly rash across the face, lupus; and recurrent jaundice, chronic active hepatitis. But clinical diagnosis of the various autoimmune diseases is rarely an uncomplicated, quick exercise. Most of these conditions fluctuate in intensity, sometimes remitting, sometimes relapsing. Often they begin as a trivial problem, a mere nuisance, readily dismissed by those affected and their doctors. The illness can creep up on you. Even as symptoms continue, or as they recur, gaining significance in daily activities, it can be difficult to specify the disease. Frequently, the complaint seems vague and unfocused, withstanding efforts to fix it as a singular disease, rebuffing any diagnostic certainty.[8]

Miriam C. Chellingsworth recalled how, as a young doctor in Birmingham, England, she tried to ignore backache and transient numbness in her legs. Later, she shrugged off the numbness on one side of her face. Eventually, though, she became blind in one eye. "I now realized," she wrote, "that I had MS."[9] Always clumsy, New Zealander John Brown at first gave no heed to his stumbling and tripping; but after a few years he consulted his family doctor, who tried to reassure him, to no avail. "My walking became more awkward; my eyes began to 'jump'; the dizzy spells increased in intensity." Five years after the start of his symptoms, a neurologist examined him and diagnosed multiple sclerosis.[10] In 1975, a nursing student was running late for a class in St. Paul, Minnesota: "I felt my leg give out. . . . I told myself I was hurrying too fast and was overly tired." A few days later, she consulted a physician after experiencing difficulty walking. He told her it probably was multiple sclerosis. "I drove home in a state of shock, crying so hard that I wonder how I arrived there at all." Before long, she was lamenting "how rudely this disease had taken over my life and . . . totally ruined the plans I had made."[11]

"There had developed a stutter in my gait, sometimes, though not always, when I wanted to step onto a pavement," wrote British novelist Brigid Brophy. "I could find no correlation with any particular circumstance." Soon she regarded household chores as "perilous," and even resorted to crawling around the house.[12] She avoided stairs and used a tightly furled umbrella to help with walking. One physician suspected arthritis; another wondered whether she was drinking too much. For a few years, doctors treated her thyroid deficiency, confident her mobility would improve as a consequence. "The trail the doctors pursued first was a false one," Brophy observed ruefully. Eventually she was admitted to hospital, where she received a diagnosis of multiple sclerosis. "The psychological and many of the social constructs through which I had been negotiating my environment trembled. Chunks of my history threatened to tumble into the limbo of non-valid concepts."[13]

In Melbourne, the patients of Ian Mackay on the Clinical Research Unit of the Hall Institute, mostly working-class people with liver disease, frequently experienced similarly vague and stuttering onsets of autoimmune disease. Eventually a number of scattered ailments and nuisances were consolidated into a distinct chronic illness and a new life course. In 1963, a forty-two-year-old woman recalled that six months earlier, as she traveled through the country in her recreational vehicle, she was "feeling tired and cranky in the morning." She noticed a facial rash, and she wanted to throw up when she thought of food. After a few months, her husband thought the whites of her eyes had become yellowish. The local doctor attributed this to infectious hepatitis, but at the end of the year she was still yellow and, according to Mackay's investigations, positive for LE cells, which suggested autoimmune liver disease. A sixteen-year-old salesgirl came to the clinic complaining that she had not been able to shake off a cold for the past seven months. She was tired and off her food and for weeks her skin had been yellowish. She too received the diagnosis of autoimmune hepatitis. A fifteen-year-old boy, a keen sportsman from the hills outside Melbourne, turned up, worried about months of general aches and pains, tiredness and nausea. The "pale, thin lad" with a yellow hue was thought initially to have unresolved infectious hepatitis, but in 1956 Carleton Gajdusek performed the first autoimmune complement fixation test on him and it proved positive. "I think that the long-term outlook is very poor," a specialist wrote to the family doctor. A thirty-seven-year-old munitions worker came to the clinic concerned about years of joint pain, headache, and nausea. For two months her urine had been exceptionally dark and her stools like "putty." This "young, thin woman" also showed LE cells and a positive complement fixation test, signaling the aggressive autoimmune disease that killed her within six years. A teenage farm boy came to the hospital after many years of intermittent diarrhea, which caused him emotional distress, and recent onset of mild jaundice. His diagnosis was ulcerative colitis and chronic hepatitis, both autoimmune. He gave up plans to train as a civil engineer. On the ward he remained "very quiet and withdrawn." He died in the following year.[14]

It seems the discovery of an autoimmune basis for a disease usually has precipitated the development of a specific serological test, which then helps to resolve diagnostic confusion. Conversely, one might argue that it was laboratory assays of immune function that made autoimmune disease conceivable in the first place. When the presence of LE cells in lupus erythematosus came to connote increased immunological reactivity against the body's tissues, the hitherto mysterious diagnostic marker additionally came to represent a once unthinkable mode of pathogenesis. When physicians in Melbourne, acting on a hunch, found that cases of chronic active hepatitis also tested positive for LE cells, these findings led to both the naming of a putative diagnostic entity, lupoid hepatitis, and the discovery of its autoimmune cause. Similarly, since the late 1940s, the presence of rheumatoid factor has distinguished rheumatoid arthritis from other joint disease and degeneration and has, at the same time, signaled the condition's autoimmune causation. In the 1950s, scientists used various complicated and time-consuming laboratory techniques, including immunofluorescence, agglutination of red blood cells, and fixation of complement, to detect antibodies, thereby amplifying their ability to recognize autoimmune processes.[15] As industry became interested in serological testing, the methods for detecting immunological function—and aberration—became miniaturized, standardized, and automated. From the late 1950s, clinical immunologists leaned heavily on immunofluorescence testing to identify antibodies to the cell nucleus (indicating systemic lupus erythematosus and other autoimmune disorders) and a wide array of specific autoantibodies, including those against smooth muscle (indicating autoimmune hepatitis), mitochondria (primary biliary cirrhosis), gastric parietal cells (pernicious anemia), thyroid epithelial cells (Hashimoto's thyroiditis), adrenal cortical cells (Addison's disease), striated muscle (myasthenia gravis), and many others.[16] Just a decade later, serologists were able to link an enzyme to antibody and substrate, giving a colored, readable endpoint. With the development of that enzyme-linked immunosorbent assay, ELISA, in 1976, the business of specific antibody detection was rendered almost routine.[17] Of all the major autoimmune diseases, only multiple sclerosis remained serologically invisible, still requiring clinical acumen and, more recently, radiological images of demyelination for its diagnosis.

Significantly, the new molecular methods of diagnosis were grafted onto older, conventional clinical entities, even though they implied or indicated a new, even generic, mechanism of disease causation. As laboratory techniques became more specific, they tended to sharpen, not challenge and overturn, generally accepted diagnostic categories. Thus, detection of diseases like rheumatoid arthritis, lupus erythematosus, and chronic active hepatitis was faster and more certain. In equal measure, knowledge of the causative mechanism of these chronic illnesses became increasingly accurate and precise, conveying a more reliable impression of their natural course and prognosis, and suggesting new treatments that involved suppression of the immune response. These diagnostic investigations must have puzzled most patients, few of whom would have imagined an autoimmune process; yet test results would set them on a different path, perhaps a strange one.

❋ ❋ ❋

"Sporadically it is, in its manifestations, a disgusting disease," Brigid Brophy wrote of her multiple sclerosis. Its unpredictable, erratic assaults on her body frustrated her, making it hard to accomplish anything, forcing her to become dependent on others. "That is the chief curse of the illness," she reflected. "I must ask constant services of the people I love most closely." As the disease worsened, leaving her permanently disabled, memories of her past abilities became more vivid. "It is an illness that inflects awareness of loss. . . . All that has happened to me is that I have in part died in advance of the total event."[18] She boasted of many minor victories, brief reassertions of personal control over her disobedient body, but gradually the medical trajectory became worse. Brophy's illness proved debilitating and refractory.

Multiple sclerosis taught Florence Lowry, a young woman when first afflicted, that life was not going to be fair. "What was most frightening in those early years was the lack of any certainty to my life," she wrote. The intermittent attacks disturbed her confidence and sense of worth. "I wanted to go back to being me but the disease had become my only role."[19] Miriam Chellingsworth reported, "The fear of losing my independence is much greater to me than fear of actual disability." She stopped making plans and setting goals. "I've learned to live one day at a time."[20] Multiple sclerosis forced many others to redefine what was "normal" for them, to accept the limitations and fickleness of the body. Anne E. Kinley, the Minnesota nursing student struck with multiple sclerosis, turned to religion for guidance and eventually came to terms with her condition. "I can honestly say that having MS is not all that bad," she wrote. "I am gaining from it the ability to live my life in a lower gear and am seeing much beauty along the way."[21] But her body was no longer reliable. She regarded it as a faulty instrument.

"You don't know what to expect from this disease or how long it will be before it gets worse," Dutch journalist Renate Rubinstein wrote of her multiple sclerosis, diagnosed in the 1970s. "No matter how you looked at it, it was defeat." She worried about appearing ridiculous and being rejected. At times she felt detached from life. "Gigantic is the word for the energy you need to spend on the body alone," she wrote. "In my movements I have become a stranger to myself." Tired and weak, she was increasingly disabled. "Time has become my capital. I'm conscious of it running out, of it lessening every minute, beyond retrieval." Toward the end, she lamented, "Of what I regarded as my true self, nothing remained."[22]

In the 1950s, Mary Howard, a middle-aged schoolteacher in England, was feeling "weary and stale and apathetic"; her joints ached and swelled and stiffened.[23] Her doctor told her she had rheumatoid arthritis, which he attributed to the menopause. He offered some potions that gave only minor symptomatic relief. "The fact that the arthritis was spreading to more and more of my joints was regarded by the doctor with what I thought was undue complacency, and I took a dismal view of the matter myself," Howard wrote. The disease "was rather spasmodic in its effects. Sometimes, for instance, my knee would become very swollen and painful, and at other times a shoulder, elbow, or a toe would be the chief trouble."[24] Her hands and wrists became permanently sore and stiff. Another physician suspected focal infection and tried several courses of the new antibiotics, to no avail. A lowering diet and absolute rest seemed to help. Later, gold injections led to further improvement. "The fact is, as we all realize sooner or later," Howard concluded, "life must hold some suffering or sorrow for all of us, and would surely fail in its purpose if it did not."[25]

When her rheumatoid arthritis was diagnosed, Mary Lowenthal Felstiner was a twenty-seven-year-old history graduate student at Stanford, in the 1960s. "Before this disease, I'd always sped forward, pushed by desire, will, and pride (I never thought of joints) through college and halfway through grad school, heading toward history teaching, raising a baby, too," she wrote. "I'd jumped into the women's movement and helped create a feminist icon—the invincible, independent woman—that I couldn't imitate after arthritis set in." Before long, "producing a meal, a memo, let alone an actual paragraph [meant] measuring out every move."[26] Physical therapy and aspirin gave little relief. For years she tried not to think about her arthritis, focusing instead on intellectual matters and political activism, but fatigue and pain limited her contributions to the women's movement. "I feel hobbled now, when I should be hitting my stride," she recorded in the 1970s. "At Passover I believed bondage couldn't last. Now I realize what doesn't last: remission."[27] In the early 2000s, Felstiner decided to break her silence. "I have nothing to look forward to but a steady downward spiral," she wrote, "yet slowly I'm arriving at another view, a view that harm is done, over time, by leaving illnesses unspoken, and that good might come of knowing the ingenuity it takes to deal with them."[28] Belatedly, she began her memoir.

One of Mackay's patients in Melbourne, a sixteen-year-old girl, took badly her diagnosis of autoimmune chronic active hepatitis. She was, he wrote in the case file in 1962, "a bright, quite intelligent youngster, . . . finishing off at school and wants to continue and learn typing." But after hearing her life sentence, she became "withdrawn, won't go out—sensitive about being 'looked at.'" She would spend the following years managing her recurrent jaundice and persistent headaches and nausea. It proved hard, too, for a fifty-year-old garage mechanic from Warrnambool to get used to the knowledge that his illness would never leave him. After receiving the diagnosis of lupoid hepatitis and rheumatoid arthritis, he told a social worker in 1965, "I am bewildered to say the least, with so many loose ends to tie up, and if I cannot work again this is the bitterest blow of all." In particular, he would always be worried that his liver "could blow up again." "I am not looking for sympathy in any way at all," he said, "my only request and hope being that my liver be put into a condition that will enable me to take up the fight for existence once more." With new medication, he survived another twenty years, but he had to give up his job.[29]

Confrontation with autoimmune disease transforms relationships and reshapes families. The needs of the chronically ill and disabled test partners, relatives, and friends, sometimes attenuating bonds, at other times strengthening them. Long-term sickness can impede and break intimacies, or deepen them. Many of the younger autoimmune patients we have known feared that their disease would prevent their finding a life partner. Women have often wondered if the illness meant they could not become pregnant and raise children. Those with offspring have doubted they would live long enough to see sons and daughters grow up. They wondered how their children would cope without them. Every one of them feared burdening loved ones. Some asked if their disease signaled a familial predisposition to this sort of condition. A few families did seem prone to immunological self-reactivity of different sorts, and occasionally a patient would act as a sentinel for more widespread illness among relatives. Rarely, one finding might start a diagnostic cascade, changing a family forever.[30] Communication with the chronically ill by medical personnel meant bearing in mind their families and their desires.

The perception of bodily failure or unreliability informs the lives of most sufferers from chronic illness. "The trajectory of chronic illness assimilates to a life course," observes Arthur Kleinman, "contributing so intimately to the development of a particular life that illness becomes inseparable from life history."[31] Chronic illness can induce fear and uncertainty, loss of control, limitation of activities, and social isolation—all factors contributing to the unsettling of identity. The sick person struggles to salvage a self from this mess.[32] There is a pressing need for what Juliet Corbin and Anselm L. Strauss call "biographical work," in which body and self can gain new meanings. The work includes incorporation of the illness trajectory into the biography, acceptance of consequences of failed or deferred performance, the reconstitution of identity, and the tracing of new directions in life, even as one grieves for what is lost.[33] Over time, one's sense of self comes to accommodate the illness.

Time is of the essence, however. In chronic illness—and autoimmune disease epitomizes this—one can be left with the feeling of a foreclosed future and a lost past, the person left behind. The present may seem unbearably urgent and short, or incredibly slow and boring. Time consumed by illness management might limit time devoted to personal projects, the time given to reconstituting a salvaged self. Time can go too fast and too slow; it can be too little and too much. The time of the chronically ill differs from the time of the well. Nor is the time of the patient the same as the time of the physician and the scientist. The sick know this in ways their caregivers scarcely fathom.

❋ ❋ ❋

On April 20, 1949, Philip S. Hench, a rheumatologist at the Mayo Clinic in Rochester, Minnesota, told colleagues of striking results he had achieved in experiments with compound E, or cortisone.[34] For some years, patterns of remission in rheumatoid arthritis had intrigued Hench, ever since he observed the diminution of symptoms in a jaundiced patient. After that, he attempted to induce jaundice in other cases of advanced rheumatoid arthritis, hoping the "stress" might somehow alleviate the joint pain and prevent further deformity. In 1948, the Mayo Clinic had admitted an arthritic young woman who was willing to try anything to gain relief. When Hench failed to induce jaundice in her, she refused to leave. Desperate, the brash, assertive physician recalled a conversation with his colleague Edward C. Kendall, a biochemist who had isolated some crystals, called compound E, from the cortex or shell of the adrenal gland. The function of this material, a corticosteroid, was still obscure. The pharmaceutical firm Merck and Company had become involved in refining the mysterious substance, assuming it might have some military uses, but it proved very difficult and expensive to isolate even a small quantity. Kendall grudgingly gave Hench some of his stock. Hench speculatively injected his demanding patient with the cortisone twice a day, and within forty-eight hours he found her in a state of euphoria, telling everyone her symptoms had disappeared. Once bedridden, she now wanted to dance around the hospital ward.[35]

Hench and colleagues approached Merck in a bid for samples of the scarce compound. Reluctantly, the company doled out a little more, enough for a limited trial. Its doubting medical director decided to visit Rochester on a chilly morning in the fall of 1948 to check on the "miracles" supposedly occurring there. According to Hench's colleagues, the company man's "icy detachment and frigid skepticism rivaled the temperature at that time." He examined critically and dismissively a young man with severe rheumatoid arthritis who claimed the injections had taken away his pain and stiffness. "Doc, you may be a good doc where you come from," the blunt Midwesterner told the man from Merck, "but as far as I'm concerned, you're full of shit." The Mayo physicians thought this "forthright retort" shook the medical director's confidence.[36] In any case, he returned east converted, eager to invest in the wonder drug.

Cortisone gave dramatic relief to several other patients at the Mayo Clinic, including those with a range of allergic or "collagen" diseases, but soon it became clear that they would relapse when the drug was withdrawn. The mechanism of action still was puzzling, though it seemed to be acting pharmacologically rather than compensating for any physiological abnormality, such as adrenal insufficiency. As Hench reported, "in an unknown manner they [the cortisone injections] provide the susceptible tissues with a shield-like buffer or protection against a wide variety of irritants."[37] Even as rheumatoid and other patients became dependent on the new drug, with its marvelous if recondite contrivances, they began to suffer manifold side effects. They gained weight and developed a "moon face," their skin thinned and hair became dry and brittle, and acne often reappeared. Some patients fell into depression; others soared into euphoria. Later studies showed that those on cortisone for a long term were prone also to high blood pressure, diabetes mellitus, opportunistic infections, and thinning of the bones, or osteoporosis.[38] They were experiencing high levels of cortisol in the blood over time, which made them "cushingoid," displaying this particular complex of symptoms.[39] This stimulus to additional disease—sometimes to fatal disease—represented the dark side of corticosteroid treatment.

At first, physicians struggled to obtain the new drug. Their patients clamored for it, but Merck could manufacture only a few grams at a time. In 1949, the U.S. National Academy of Sciences (NAS) established an expert committee to allocate cortisone among accredited researchers, so they could investigate its properties, following the model of penicillin rationing during World War II. Officially, cortisone would be available only in valid clinical trials. Rheumatologists like Hench were furious, arguing that science bureaucrats were preventing them from relieving the suffering of their patients. They wanted to use cortisone pragmatically, just like any other drug. Now enthusiastic about its property, Merck feared losing control of the clinical research and missing out on future marketing and distribution opportunities. Before long, the black market in cortisone thrived. By the end of 1949, the NAS disbanded the committee, as Merck managed to gear up production.[40] Within a few years, Upjohn scientists discovered a mold that converted the hormone progesterone to cortisone—and progesterone was plentiful, since Syntex had found a means of isolating it from the Mexican yam, *barbasco*.[41] Soon prednisone or prednisolone joined cortisone on the burgeoning, worldwide, corticosteroid market.

"The therapy of autoimmune disease is based on the intelligent use of the corticosteroid drugs," Mackay and Burnet advised in 1963.[42] During the previous decade, various informal trials had demonstrated that prednisone and other corticosteroids suppressed immunological reactivity and limited tissue damage from the interaction of antibodies and antigens. It now seemed possible that patients who were once expected to die within a few years could approach a normal life expectancy. It was convenient, even propitious, for clinical immunology that the corticosteroids, with their power to damp down the immune system, had come along just when autoimmune pathogenesis was recognized. The new drug class added an effective therapeutic armamentarium to the specialty's prolific diagnostic technologies. But some autoimmune diseases would prove more susceptible to corticosteroids than others: hemolytic anemia, lupus erythematosus, chronic hepatitis, and rheumatoid arthritis were more readily contained than multiple sclerosis.[43] Moreover, the new drugs did not cure autoimmune disease. They were no panacea. The requisite long-term maintenance therapy usually brought other health problems, and sometimes physicians found it necessary to seek alternative "steroid-sparing" treatments. Ionizing radiation could inhibit the immune response, but as

Mackay and Burnet averred, "total body irradiation is clearly not a practical approach to the elimination of committed immunologically competent cells, whether normal or forbidden, from the body."[44] Occasionally the excision of lymphoid tissue, such as the spleen, where white blood cells or lymphocytes multiplied, could reduce immunological pathology. But the most common practice in the 1960s was to substitute, as much as possible, cytotoxic agents, which inhibited cell proliferation, for corticosteroids, thus restricting the side effects.

"Incidents"—or rather, unforeseen disasters—during World War II had revealed that nitrogen mustard, or mustard gas, could destroy white blood cells, often with fatal effect.[45] Once made public, these findings caused scientists to ponder whether similar, though less deadly, substances might target malignant or cancerous white blood cells—that is, combat leukemia. After the war, George H. Hitchings and Gertrude B. Elion at the Burroughs Wellcome laboratories in New York tried to find some "decoy" molecules that cancer cells would take up in error and incorporate into their structure, thereby killing them or impeding their proliferation. They focused on identifying a misleading substitute for one of the constituents of DNA, since it was known to be a crucial component in cell division or reproduction and thought probably responsible for heredity. In 1951, they made a purine analog, 6-mercaptopurine (6-MP), which blocked DNA synthesis and stopped cell multiplication.[46] Given to children with leukemia, it sometimes produced temporary remissions in the disease. Other antimetabolites or cytotoxins also offered promise in the fight against cancer after the war. Scientists at Lederle developed aminopterin, which gave cells the false impression of being folate, another molecule needed for their replacement and proliferation. Boston pathologist Sidney Farber showed that children with some forms of leukemia also benefited from this drug, despite its general toxicity.[47] Methotrexate, a less harmful version of aminopterin, came on the market in 1956; and the more palatable azathioprine (Imuran) largely took the place of 6-MP in the early 1960s. A series of clinical trials and laboratory studies in this period indicated that both azathioprine and methotrexate additionally suppressed the rapidly dividing cells of the immune system, thus reducing immunological reactivity and making tissue transplantation feasible.[48] Soon azathioprine, in particular, was contributing to the treatment of autoimmune disease, often as a steroid-sparing agent, more rarely as a first option.

While serological diagnosis of autoimmune disorders had become ever more specific, suppression of the diseases remained tellingly general and holistic. Certainly, some sorts of autoimmune disease appeared especially responsive to one drug or another: lupus erythematosus and chronic hepatitis usually reacted to azathioprine; rheumatoid arthritis often was susceptible to methotrexate. But the basic principle of suppression of self-reactivity was fundamentally physiological or constitutional—even symptomatic—and not specific. The medical aim was to reset the body's homeostatic mechanism, to restore its immunological equilibrium, to reinstate the normal condition of self-tolerance. Although physicians could look for laboratory evidence of tissue damage, they continued to rely primarily on their patients' symptoms to evaluate disease activity and to guide their selection of treatment options. They tended to titrate the corticosteroid dose against individual expressions of well-being and illness. Their efforts to depress or stimulate the patient's immune responsiveness—to rebalance the system—therefore resemble the constitutional medicine of the early nineteenth century, with its concern for bodily sensitivity, integrity, and idiosyncrasy. As Burnet put it, "the actual facts of autoimmune disease are so individual and only broadly reproducible that any approach must of necessity be a flexible one."[49] Thus, modern therapy for autoimmune disease emerged anachronistically as a biographical therapeutics, obsessively modulating self-reactivity, persistently adjusting medication to influence self-tolerance. It was a deeply felt, closely experienced, therapeutics—in contrast to the covert operations of standardized immunological manipulations, such as vaccination and antibiotic medication.

"Every time something new is invented I get in on the ground floor with it," Flannery O'Connor assured a friend in 1957. "There have been fine improvements in the medicine in the seven years I've had the lupus."[50] Maintained on prednisone, the writer's symptoms were manageable for many months at a time, but other serious problems gradually accumulated. "What they found out in the hospital," O'Connor wrote in 1960, "is that my bone disintegration is being caused by the steroid drugs which I have been taking for ten years to keep the lupus under control. So they are going to withdraw the steroids and see if I can get along without them."[51] But corticosteroids, so damaging to her bones, were the only drugs that adequately suppressed her lupus. The therapeutic impasse made her last years particularly miserable.

Mackay's autoimmune patients frequently felt the effects of their medications, especially the corticosteroids. "I feel as fit as a trout, but I'm still gaining weight," a married woman in her late fifties complained in 1964. After years of prednisone she had a moon face, bad skin, and constant headaches. She accepted that she was irritable and intolerant, just "marking time." Others became tearful and depressed; some experienced a destructive and profligate euphoria. There were a few who became mistrustful and paranoid, and several contemplated suicide. A thirty-year-old woman from the country, suffering from ulcerative colitis and chronic hepatitis, developed osteoporosis and endured frequent bone fractures. She was lethargic and sleepless, with "nothing to live for"—she wondered if she was going insane. Young men and women often fretted about changes in their appearance. One moon-faced sixteen-year-old girl was "extremely distressed about change in facial appearance resulting from prednisolone," Mackay noted in 1963. He suspected she was, consequently, "forgetting" to take her medication, like many other patients. A fifty-year-old woman, caring for her invalid mother, hated her moon face and lamented her weight gain and bruising. Her back was giving problems, cuts and sores would not heal, and her legs had become so weak she could not climb stairs. But every time she stopped the loathsome steroids her lupoid hepatitis returned. When the effects of treatment became dangerous or exceptionally distressing, Mackay tried to substitute azathioprine (or more rarely 6-MP) for prednisone, but the autoimmune disease often could not be contained on the new regimen. "I seem to be going down hill a bit fast for comfort and the corners are a bit sharp," one patient wrote to Mackay after a change in medication. A flexible combination of immunosuppressives, however, usually reduced the side effects of each member of the therapeutic ensemble. Eventually an acceptable arrangement would be negotiated, a balance achieved—then superseded, then another bargain struck, then revised—and so life went on, often for many more decades than anyone could have envisaged before the advent of these medications.[52]

❋ ❋ ❋

Reviewing Mackay's case files from the 1950s and early 1960s, we were surprised by the high proportion of autoimmune patients mentioning close relatives—siblings, children, parents, uncles, and aunts—with juvenile-onset diabetes mellitus. No one had noticed this association at the time, though some clinical immunologists had begun to speculate on possible autoimmune explanations for that disease. After considerable discussion in 1963, Mackay and Burnet decided to omit juvenile, or type 1, diabetes mellitus from the register of autoimmune diseases in their textbook. "The climate of opinion on autoimmunity in 1963 seemed too inhibitory to consider the inclusion of that disease!" Mackay later reflected.[53] Within a few years, however, he would regret the lapse.

Diabetes had long been recognized as a heterogeneous disease, with distinct manifestations. Even in the 1880s, Parisian physician Étienne Lancereaux distinguished two body types among those afflicted: *diabète maigre*, generally the scrawny young, who often succumbed within months; and *diabète gras*, stout persons diagnosed in middle age, who lingered with the ailment.[54] In the 1940s, New York enthusiasts for the fashion of somatotyping formally reiterated the morphological distinction: type 1 diabetics were thin, and type 2 large.[55] With the introduction of pharmaceutical insulin in the 1920s, physicians observed that the thinner group, whose inability to control blood sugar levels often declared itself early in life, demonstrated a dramatic sensitivity to the new drug. Regular insulin injections allowed them to lead normal lives, but if they missed doses, the symptoms and signs of diabetes rapidly returned. Their insulin deficiency made them dependent on the replacement therapy. In contrast, overweight patients affected in middle age tended not to respond to insulin. Rather, they managed their diet and carried on until the pathological consequences of sustained high blood sugar levels eventually caught up with them. For most of the twentieth century, diabetics were sorted into these morphological and secular classes.[56] Most physicians told their diabetic patients that the cause of the condition was still unknown, probably the result of genetic predisposition and diet. A few epidemiologists discerned links to previous viral infection, especially mumps and intestinal disease, but evidence was scanty.[57]

In the nineteenth century, Lancereaux recognized that the pancreas was at fault in diabetes, but the nature of the lesion resisted simple explanation. The process in early-onset diabetes must involve damage to the islet cells of the pancreas, which produce insulin, it was thought, but pathologists rarely observed any inflammation at autopsy. In general, they were looking at only the very late stages of the disease, long after the initial mischief had occurred. But in the 1960s, pathologists examining specimens from patients with type 1 diabetes who had died soon after diagnosis frequently saw insulitis, or inflammatory changes around the islet cells.[58] They took to speculating on the cause. Recently, investigators had induced insulitis in cows injected with bovine and porcine insulin mixed with Freund's adjuvant; another scientific team achieved the same result in rats receiving anti-insulin serum.[59] Perhaps, therefore, the pancreatic lesions in type 1, juvenile-onset, insulin-dependent diabetes had an immunological origin. Could this be another autoimmune disease? Efforts to identify autoantibodies to islet cells frustrated the scientists until 1974, when they began to use fresh, unfixed human pancreas tissue rather than conventional rodent samples. Suddenly, the antigen-antibody complexes glowed under the florescent microscope.[60]

The discovery of the autoimmune cause of type 1 diabetes in the 1960s and 1970s expanded the repertoire of clinical immunology, permitting it to encompass a larger pediatric population and to broaden its explanatory scope. Moreover, it offered new laboratory tests to refine and secure the distinction between type 1 (autoimmune) and type 2 (metabolic) diabetes. Unfortunately, immunosuppressive medication would be pointless, since diabetes emerges only at the terminal phase of autoimmune insulitis, when the immunological damage already has occurred. Discovery of the autoimmune cause in this case failed to alter the disease's natural course and treatment. Those afflicted with type 1 diabetes continued to understand it as a disease of insulin deficiency—rarely did they consider it an autoimmune phenomenon.

Type 1 diabetes generally is experienced as a particularly onerous, long-term illness, one requiring daily insulin injections, regulation of diet and exercise, and monitoring of sugar levels. "It's a case of you've got to have [insulin] or you die," an elderly Englishwoman, reflecting on seventy-seven years of insulin, told an interviewer. "It's life for you, injections, when you're diabetic." A woman from Yorkshire, diabetic for more than sixty-five years, remarked: "I think the principal effect it's had is that it's meant I've always had to be rather precise in what I do and how I think of things. I have to lead an ordered life and know what the next step is going to be, so I can make preparations for it." She learned to build her life around insulin. "I was always afraid that it would rule my life," said a London physician, a meticulous insulin injector, recalling more than forty years of diabetes, "and I was always pretty determined that it shouldn't. I mean, it's all very well to have a chronic disease, but you want to control the chronic disease, you don't want the chronic disease to control you."[61] Some sufferers tried to deny their illness altogether. Aged sixteen, excessively thirsty, and losing weight, Peter Corris was diagnosed with type 1 diabetes as a student in Melbourne. The future historian and novelist imagined diabetics as "flawed, frail creatures," usually impotent—not like him. Although he injected insulin regularly, Corris chose to ignore his diet, to smoke, and to drink heavily. "I believe I was ashamed of being a diabetic," he wrote.[62] When he started to lose his sight, he gave up smoking, though he continued to drink too much alcohol. The shock of talking with Fred Hollows, the distinguished Australian eye surgeon, turned him into a more "responsible" diabetic. "What's this gut you've got on you?" Hollows rasped bluntly. "You've had all this expensive treatment and you're just fucking throwing it away. . . . If you don't get off the piss [alcohol] and get fit you'll be blind in five years and dead in ten."[63] Chastened, Corris gave up his "bravado" and learned to live with his chronic illness.

From the late 1950s, as autoimmunity became more popular as an investigatory path; it served as a potential explanation for any number of mysterious diseases, especially emerging ones, though rarely was the putative etiology validated as in type 1 diabetes. When D. Carleton Gajdusek, having recently discovered the autoimmune complement fixation test, visited the Fore people in the eastern highlands of New Guinea in the late 1950s, he wondered if kuru, the fatal brain disease afflicting these people, might be autoimmune in origin. Like everyone else, he knew the Fore as cannibals, ritual consumers of their dead loved ones, and it seemed feasible that eating others' brains might give rise to antibodies against human brain tissue. But at autopsy, the kuru dead showed no sign of inflammation in their brains, so Gajdusek abandoned the idea in favor of some vague genetic hypothesis.[64] Yet, autoimmunity remained an appealing explanation for many other frightening emergent diseases. The recognition of acquired immunodeficiency syndrome (AIDS) in the early 1980s led some medical scientists to speculate on a possible autoimmune causation. Perhaps the immune system was attacking itself, with its hyperactivity disclosed paradoxically as deficiency. But why would so many young men suddenly be developing autoimmune disease? Even after discovery in 1983 of what came to be called the human immunodeficiency virus (HIV), which infects and destroys certain lymphocytes, some scientists continued to wonder if the virus might be triggering an autoimmune response, a reaction of the host against its own immune system.[65] In the early 1980s, for a brief period, clinical immunologists seized on AIDS as an area for potentially fruitful work, claiming clinical authority, before infectious disease experts properly took over management of the disease.[66] Thus, type

1 diabetes remains one of the few major illnesses to secure an undisputed autoimmune explanation since the 1960s.

❋ ❋ ❋

Research in immunology was a minority interest among ordinary physicians through the Cold War period, even as the specialty of clinical immunology matured and secured its grip on major academic medical centers and teaching hospitals. Clinical immunologists, though few in number, proved adroit in making connections between patient management and the sequestered procedures of the scientific laboratory. Their clinical expertise crossed traditional boundaries, permeating pediatrics, adolescent health, internal medicine, and geriatrics. The new specialists boasted uncommon skills in immunological diagnosis and immunosuppressive therapeutics, and a vital ability to perceive, monitor, and regulate the pathological self-reactivity that triggered a variety of diseases. Although they could not eliminate autoimmunity, they usually could damp it down and control its manifestations, often for many decades. The laboratory connections of the clinicians thus altered the illness trajectories of those afflicted with autoimmune diseases, modifying their experiences of body, self, and time. Additionally, the clinical connections of the physician-scientists suggested fresh problems and materials for laboratory research, different routes for immunological investigation, and new perspectives on human reactivity and sensitivity.[67] Even so, long-lasting engagement and close encounters with patients meant that physician-scientists found it hard to regard these sick people simply as experimental subjects, as mere research prompts.[68] These connections were humanizing immunological research, spoiling any ideals of detached biological investigation.[69]

Few ordinary people felt a pressing need to understand the concept of autoimmunity; accordingly, few patients perceived their disease as autoimmune. On the whole, there seems to have been a poignant failure in the clinic to communicate new, admittedly somewhat esoteric, biomedical concepts during this period—despite the public ambitions of Burnet and other scientists. It could be argued that, while the practices of clinical immunology dominated the life world of many patients, the specialty lacked intellectual hegemony—or even a commitment to discussing with lay people the ideas on which the field was based, its creed. Clinicians and patients were all doing biographical work, trying to refigure or reconstitute the self, but the two groups followed different paths that rarely intersected. Clinicians sought to restore the integrity of the body, to lessen reactivity to self, by suppressing immune responses. Patients tried through social means to restore a sense of self, to reclaim or reconstitute a self that was displaced by chronic illness and disability. Yet, these disparate concepts of self—one disease based, the other concerning disability, equally legitimate but functionally distinct—hardly ever made contact, let alone interacted. Perhaps the clinical conventions of the era, regardless of the sympathy of doctor and patient, interdicted conversations of that sort.

Over time, however, autoimmunity has become a more useful conceptual tool for some medically literate patients. In 1991, Virginia T. Ladd established the American Autoimmune Related Diseases Association (AARDA), a self-help and advocacy group dedicated to promoting research and education.[70] Suffering from systemic lupus erythematosus, Ladd had been a founding member of the Lupus Foundation, but she soon realized that many other diseases shared its pathogenic terrain. From the beginning, Ladd made sure to include in the association's membership leading scientists and clinicians, experts on autoimmunity, who contributed to research advocacy without dominating the association's activities. Later, AARDA pressed for and succeeded in creating the National Coalition of Autoimmune Patient Groups, drawing together the older, disease-based self-help groups to lobby legislators and research bureaucrats. In response, the National Institutes of Health established a committee to coordinate all sponsored autoimmunity investigation. Early in the twenty-first century, the idea of autoimmunity was beginning to have more explanatory power for some patients. It was coming to shape their sense of themselves and to suggest new social affiliations.[71] Mary Felstiner, for example, recognized that her rheumatoid arthritis was an autoimmune disorder. For her, this meant that the immune system "keeps treating its body parts as aliens, keeps rousing immune cells until they're attacking familiars as suspects, picking on bones they're supposed to be keeping safe." She understood that "far from deficient, my teams of immune cells are so proficient they're overdoing the job. For the first time I'm impressed by that." Equally, it was a depressing thought. "The hardest lesson to learn from RA," Felstiner reflected, "is that life-preserving forces ravage us, putting an end to endless possibility."[72]

CHAPTER SIX

Reframing Self

❋ ❋ ❋

Toiling in the laboratories of the Rockefeller Institute in the late 1930s, not far from where Rivers and Schwentker were trying to induce encephalitis in monkeys, Merrill W. Chase wondered how skin could be so exquisitely sensitive to simple chemical substances. For example, a tiny amount of poison ivy elicited, after a delay, an intense reaction in his guinea pigs.[1] The mechanism of this allergy, or sensitization, puzzled Chase. In the early 1940s, his senior colleague, Karl Landsteiner, suggested an attempt to transfer the allergy from one guinea pig to another in order to find its cause. Surprisingly, blood serum, which lacks cells but would contain antibodies, showed little effect on untreated guinea pigs. In contrast, washed white blood cells or lymphocytes, presumably free of antibody, did transmit this delayed-type hypersensitivity.[2] Excitedly, Chase called over Landsteiner. "When he looked at the exudate cells under the microscope," Chase recalled, "and saw lymphocytes, he exclaimed: 'Yes, I thought so'—a gargantuan fib."[3] For years, Landsteiner had argued for a single immune response based on antibody production, and he reprimanded those who questioned the economy of nature.[4] Now, near the end of his career, he saw evidence that cells alone could instigate an allergic reaction, implying an additional immunological contrivance, a mystery to him. Chase confirmed the immunological activity of white blood cells in further experiments with killed tuberculosis germs, or the derivative tuberculin, injected into the skin, known to cause a particularly tardy allergic reaction.[5] As he reported to the Rockefeller Institute, "in actively sensitized guinea pigs, antibody in the serum can be shown to transfer an 'evanescent' type of reaction to other, previously untreated animals, while the white cells will transfer the 'delayed' or tuberculin type of reaction."[6] Irritated guinea pigs in New York thus heralded the beginning of cellular immunology. By the end of the 1950s, as we have seen, the previously disparaged small lymphocyte would become the privileged cellular actor in the immune drama, a versatile player responsible not just for antibody production but also for other activities associated with delayed-type hypersensitivity.[7]

During the 1960s, immunologists meticulously set about determining lymphocyte classes, differentiating the potencies and tracing the interactions of these defensive cells. Soon, it was commonly accepted that the immune system consisted of more than a standard retinue of antibody-producing lymphocytes. Rather, it seemed to be a complicated, self-regulatory network of variously competent cells along with antibodies and chemical factors. The idea of self-tolerance, or the absence of self-antigenicity, which so intrigued Burnet, therefore required adaptation and reformulation. Since the immune repertoire was more extensive and complex than anyone had imagined, explanations of how it recognized, or mistook, the body's own tissues demanded modification. Even as the scene of investigation was shifting inexorably from the clinic back to the research laboratory and animal house, the discovery of the population dynamics of lymphocytes also would eventually entail the rethinking and reframing of the clinical problem of autoimmune disorders, the failures of self-tolerance.

A deeper understanding of the molecular meaning of individuality accompanied the differentiation of cellular function in the immune network. Even before World War II, scientists knew that cells of mice and some other animals exhibited distinctive patterns of antigens, as markers of individuality. Discoveries in the 1950s of antigens on human white blood cells—human leukocyte antigens—seemed almost to confer molecular form and functional significance on Burnet's old, discarded theory of "self-markers." Unlike the distinctive, and long familiar, antigens on red blood cells—which Landsteiner had used for ABO blood grouping—these white blood cell antigens were shared with most of the other nucleated cells of the individual. Therefore, they offered an explanation of "histocompatibility," why some foreign tissues were more likely to be rejected than others; accordingly, they became known as the "transplantation antigens." The greater the histocompatibility or tissue similarity—that is, the better the homology of self-markers—the more likely a graft would take hold. Not until the 1970s did immunologists discern the role of these transplantation antigens in immunological recognition and response. Strangely, the tissue antigens appeared to stimulate and coordinate the lymphocytes. Through essential coupling to a foreign antigen, they allowed its detection and processing. They drew attention to anomalies in the antigenic stock of the body. They acted like guides or chaperones to the immune system, like its personal assistant, insistently whispering about the status of guests at a party. Immunologists came to realize that, as Peter C. Doherty put it, the "so-called transplantation system is, in fact, a self-surveillance system."[8]

The immunological self of the clonal selection theory had been a negative homunculus, whatever was left after the deletion of clones that reacted to the body's own tissues; whereas tissue antigens marked out a positive, even possessive, immunological self, a self continuously involved in assessing menace and likeness by surveillance. The uncovering of the role of these self-markers in the immune response would necessitate further revision of explanations of tolerance and its errors, the autoimmune diseases.

In this chapter, we consider the impact of discoveries in cellular immunology and histocompatibility on theories of autoimmunity from the 1960s until the beginning of this century. Inevitably, it is a complicated story, a partial sifting of historical worth from contemporary chaff. Not to make an effort at the first draft of this recent history, however, would endorse the self-serving claims of Burnet and Jerne that the history of immunology ended with them. The story we tell may be provisional, but certain themes do emerge with clarity. There is the irresistible rise of laboratory investigation, with its dependence on animal, particularly inbred mouse, models, its confidence in esoteric techniques, and its tendency to default to detailed molecular explanations of complex biological processes. Along with such captivating molecularization goes a decline in the intellectual salience of the clinic and the reduced influence of patients and their diseases on immunological speculation. Current vectors point from laboratory to clinic, as the new term *translational immunology* implies. Among other emergent characters of recent immunological inquiry are its vast expansion or amplification, the proliferation of immunological career paths, whether in research or clinical settings, and the dispersal and adaptation of immunological thought and practice around the world. In the past, immunological research, including studies of autoimmunity, took place successively in various major urban centers, such as Berlin, Vienna, Paris, New York, London, and Melbourne, but now it happens throughout the developed world. Late in the twentieth century, the immunological self—however ambiguous, metaphorical, and suspect—went global.

❋ ❋ ❋

In the 1950s, the thymus seemed "an enigmatic organ" to the young immunologist Jacques F. A. P. Miller. In the chest, above the heart, this "organ in search of a function" starts large in infancy but shrinks in old age to shreds of fiber and fat. Most physicians had assumed that "the thymus had become redundant during the course of evolution and was just a graveyard for dying lymphocytes."[9] Although lymphocytes were known generally to be involved in antibody production, those resident in the thymus appeared inert and senescent; removal of the organ in adults did not seem to affect the immune response. As a medical student at the University of Sydney, Miller found this obscurely quiescent organ intriguing; as a scientist at the Chester Beatty Research Institute in London, he decided to experiment with it.

Working in a wooden shack once occupied by circus animals, an hour by train from the West End, Miller wondered if removing the thymus in infancy might prevent mouse leukemia. It seemed a plausible hypothesis, since leukemia is the proliferation of white blood cells, such as those apparently dormant in the thymus, and it was known that a virus introduced to mice in the neonatal period might provoke the disease. But every time he delicately extracted the thymus from a newborn mouse, the poor animal soon either wasted away from infection or was eaten by its mother. Indeed, Miller's wife Margaret spent much of her free time "coaxing their mothers into not eating them."[10] Before long, it dawned on the scientist that the belittled thymus might actually be essential for life, at least soon after birth. "If my mice had not been raised in converted horse stables but in pathogen-free quarters," he reflected, "they may never have contracted this wasting disease and I would not have been alerted to the crucial role of the thymus in the development of the immune system."[11] At autopsy, the mice revealed low lymphocyte counts and effete lymphoid tissues. These findings prompted Miller—who had attended Peter Medawar's lectures in London—to try to graft skin onto the mice before they became too weak. "To my amazement," he recollected, "grafts from foreign strains and even rats survived."[12] In 1961, Miller concluded that the thymus must be the source of a special class of lymphocytes involved in defense against infection as well as rejection of foreign tissue—and that presumably these immunologically competent cells migrated to the rest of the body before the organ shriveled with maturity. Passing through London, Burnet visited Miller's suburban shed and discussed the experiments. Shortly thereafter, the young scientist was proposing the thymus as "the site where self-tolerance is imposed" early in life—that is, the site where Burnet's "forbidden" clones are deleted.[13]

At the University of Minnesota, pediatric immunologist Robert A. Good also took an interest in the biology of the thymus, but he was slower to publish than Miller. The Midwesterner had returned from a fellowship with Henry Kunkel at Rockefeller University, curious about inherited and acquired defects of the immune system. In 1953 he saw a man who developed in adult life an inability to produce antibodies, along with a mass in his chest, which turned out to be a benign enlargement of the thymus. Good and his colleagues therefore wondered if the mysterious organ could contribute in some way to the control of immunological processes.[14] Later, students in his laboratory decided to remove the thymus of newborn rabbits and see what happened: within six weeks the animals were "immunological cripples"—or so Good recollected.[15] When Good and his colleagues removed the thymus of neonatal mice, they found the ailing, stunted survivors readily accepted skin grafts from unrelated donors. In 1961, Good presented the findings at the annual meeting of the American Association of Immunologists. He noted that Miller had found the same results "simultaneously with our experiments, apparently completely independent of our approach, and, indeed, coming to the problem from an entirely different perspective"—and had published first.[16]

Ambling around the Midwest, Good and his colleagues became aware of the research of Bruce Glick, a poultry scientist in Ohio who was fascinated by a lymphoid organ, impressively called the bursa of Fabricius, at the posterior end of the chicken's intestinal tract. Supposedly trivial, this feature of the avian anatomy was absent from mammals. In 1952, Glick, watching his professor remove the bursa, or cloacal thymus, from a dead goose in the damp basement of the Agriculture School at Ohio State University, had the gall to ask what purpose the organ served. "Why don't you find out?" was the rejoinder.[17] Glick soon determined that the bursa grew most rapidly in the three weeks after the fowls hatched, but then declined and shrank. He removed the bursa during this critical period and watched expectantly. As another graduate student desperately needed birds in order to study the effects of immunization against Salmonella infection, Glick loaned him some chickens. Oddly, all the bursectomized fowls failed to produce antibodies: and so the investigators grasped that they inadvertently had discovered the function of the bursa of Fabricius.[18] As Glick piously reflected, "the bursa's secret would still be held by the chicken if truth were not the universities' main goal."[19] Good heard of a zoologist in Madison, Wisconsin, repeating the experiments, so he sent a graduate student to take a look. His student "came back from Wisconsin full of enthusiasm and launched into experiments to test the capacity of chickens bursectomized in the newly hatched period to express themselves immunologically."[20] Even though Good and his coworkers observed that the bursectomized chickens retained the capacity to reject foreign grafts, unlike the thymectomized mice, they believed that the bursa and the thymus probably were immunologically equivalent, performing the same function.

After meeting with Miller and learning from Good about Glick's research, Burnet persuaded a couple of colleagues in Melbourne to experiment on the chicken bursa and thymus. Inoculating eggs with testosterone, Noel L. Warner and Alex Szenberg could, through hormonal means, prevent later bursa development in the chickens—while usually leaving the thymus and spleen intact—and cause impaired antibody production in their poultry. However, if the thymus was excised, the chickens failed to reject foreign grafts or develop the delayed hypersensitivity that Chase had observed—their cellular, or non-antibody-dependent, immunity was enfeebled. The scientists inferred that the lymphocytes originating in the bursa must be responsible for antibody production and those from the thymus accountable for cellular immunity—cytotoxic activity—against, for example, transplanted tissue.[21] Of course, this was all very well for fowl, but which part of a mammal might assume the role of the bursa? By the late 1960s, most immunologists supposed it must be the lymphocyte-generating tissue in the bone marrow. Increasingly, they conjectured that the two subsets of lymphocytes perform different functions yet work closely together.[22] In Melbourne, Miller and Graham F. Mitchell, an ebullient graduate student, managed to remove the thymus from adult mice, irradiated them sufficiently to strip them of their immunological capacity, and then protected them from the lethal effects of irradiation by injecting bone marrow cells. Subsequently, the immunologists injected the poor mice with a pure population of thymus-derived lymphocytes, obtained from normal mouse donors. They showed that the bone marrow had supplied the antibody-producing cells, but this lymphocyte subset required thymus-derived cells in order to mount a response to antigens.[23] Thereby, Miller and Mitchell established that thymic lymphocytes could act as "helpers" or regulators of the bone marrow–derived lymphocytes, enhancing their ability to produce antibody. Just how each clonally selected lymphocyte, with its own specificity, located its affine and communicated with it still was unsettled. Indeed, Miller recalled, "Our proof of the existence of two totally separate lymphocyte subsets in mice was greeted with surprise and scepticism by most immunologists."[24]

Yet, the following year, Ivan M. Roitt and his colleagues declared in an influential review of the population dynamics of lymphocytes that the cells indisputably sorted into two distinct classes, the thymic-dependent T lymphocytes and the thymic-independent, or bursa-equivalent, B lymphocytes.[25] ("Why Jaq and I missed this opportunity to come up with the terms T and B cells is a source of some anguish," Mitchell later admitted.[26]) According to Roitt and his team, both classes contained antigen-sensitive cells. When B lymphocytes encountered the appropriate antigen, they proliferated, producing specific humoral antibody. In contrast, Roitt regarded the long-lasting T lymphocytes as both the repository of immunological memory and as potential killer cells, cytotoxic for grafted tissue; they were contributors to delayed hypersensitivity and helpers in amplifying the antibody production of B cells. Therefore, extirpating the thymus of newborn animals would prevent graft rejection and delayed hypersensitivity and impair the antibody response to at least some antigens. The British scientists even wondered if anomalies in the T cell population might cause organ-specific autoimmune disorders, such as thyroiditis and gastritis, and B cell defects might result in diffuse self-reactivity, as in the so-called connective-tissue diseases, like systemic lupus erythematosus and rheumatoid arthritis—but these speculations turned out too neat to be biologically valid.[27]

The notion of the T lymphocyte as a monitory cell was gaining currency. In 1970, Richard K. Gershon, an iconoclastic Yale immunologist, introduced the idea that T cells could suppress as much as stimulate the immune response, thereby leading to "peripheral tolerance," a concept that harked back to Burnet's "homeostatic process." The idea that self-tolerance derived mainly from the deletion of forbidden clones in the thymus frequently was set aside, though never banished. Gershon and colleagues demonstrated that the T lymphocytes of mice that had been rendered tolerant to sheep red blood cells could transmit the same tolerance to naïve mice. They believed "suppressor" T cells to be responsible for this surprising "infectious tolerance."[28] No doubt it was an unusually convenient and appealing theory. Perhaps autoimmune disease and allergy were the results, then, of insufficient numbers or inefficiency of suppressor cells. For a decade or more, immunologists talked confidently about the properties and functions of suppressor T cells, but lymphocytes of this subset proved exceptionally elusive. No one could quite pin them down—or at least distinguish them from the cytotoxic T cells. Nor could anyone establish the mechanism of suppression. In 1983, plainspoken immunologist Melvin Cohn wondered "to what extent this suppressor pathway will turn out to be a laboratory construct of no physiological significance."[29] Interest in suppressors became more restrained, but the theory proved too attractive to be brushed aside for long. During the 1980s, various findings again suggested that some T lymphocytes actually could modulate the responses of other T cells and of B cells, though they were now reassuringly renamed "regulatory," not suppressor, T lymphocytes. Researching industriously in Kyoto, Japan, Shimon Sakaguchi found that some thymectomized mice, deficient in lymphocytes, developed typical autoimmune diseases. Then, by infusion of normal T lymphocytes, he could suppress self-reactivity and rescue the thymus-deprived mice.[30] Such regulatory T cells—familiar to the new generation of immunologists as "T regs"—thus became part of the understanding of normal immune response, maintaining "the homeostatic equilibrium of immunity and tolerance."[31] All the same, their specificity, mechanism of action, likely groups, and therapeutic potential remained open to further inquiry.

During this period, immunologists witnessed a veritable explosion in the population of cells discovered to be contributors to the immune response. As lymphocytes were differentiated into ever more refined and intricate functional subsets, the description of the role of other cells in presenting antigens to them was expanding rapidly. Evidently, infected cells processed and displayed fragments of viral antigens on the surface membrane; similarly, circulating macrophages consumed and then presented antigens for the delectation of some clone of lymphocytes. After 1973, sparsely distributed dendritic cells came to assume the major responsibility for antigen presentation and, hence, for activation of lymphocytes.[32] It turns out that these previously obscure cells, scattered throughout the body, act as immunological sentinels, alert for pathogens and other intruders, ready to digest and exhibit antigens. These "innate"—rather than adaptive—immune cells provide a readable inventory of the body's antigenic environment, a nonspecific recognition of foreign antigenic pattern that prompts specific lymphocyte activation.

Writing at the end of the century, Noel R. Rose attempted to survey the implications for understanding self-tolerance of the new knowledge of the behavior of lymphocytes and other immunologic cells. He did not doubt that self-tolerance was "based less on the deletion of potentially self-reactive clones than on the balance between clones that promote and retard harmful autoimmune processes." Autoimmunity was common, but it rarely flared into autoimmune disease, mostly because of the homeostatic mechanisms. Although deletion in the thymus of some lymphocytes excessively reactive to self must occur, it now seemed more limited in scope.[33] Therefore, self-reactive T lymphocytes—albeit of low affinity—exited the thymus and circulated through the body; shaping the immunological self becomes a lifetime commitment. These self-reactive T cells failed, in general, to produce autoimmune disease because they encountered insufficient antigenic stimulation, or they tussled with the regulatory T lymphocytes, which modulate their activities. "The ratio of T cell populations promoting and hindering the immune response directly, or through production of mediators, best describes our understanding of self-tolerance and autoimmunity." Thus did Rose elevate the thymus to the seat of "the immunological soul."[34]

But what about the humbled B cell? During the previous decade, it had become clear that B lymphocytes were less tolerant of the body than were T lymphocytes: it is normal for the former to produce autoantibodies. Indeed, most of the "preimmune repertoire" of natural antibodies—the antibodies normally present in the body without an antigenic challenge—might be autoantibodies, but their affinities often were low.[35] As these autoantibodies frequently reacted additionally, and perhaps more efficiently, with foreign substances, their *self*-reactivity could be a minor, though still vital, property. They were mundane, underperforming antibodies, held back, or controlled homeostatically, by regulatory T cell suppression, helper T cell indifference, and perhaps even by antibodies against the receptors of self-reactive B cells. A background of autoimmunity, then, was normal; yet on occasion, it would shade into overt pathology.

❋ ❋ ❋

Running parallel with studies of lymphocyte population dynamics, investigation of human tissue antigens was revealing the molecular traces of individuality. According to Jean Dausset, one of the pioneers of the field, these markers of histocompatibility, or transplantation antigens, contributed to "our growing understanding of the molecular mechanisms of individuality, of recognition of self and non-self, and consequently of the survival of the species."[36] At first, "attention was centered on the almost botanical description of their genetic polymorphism," their seemingly infinite variation of form and pattern. Soon, however, scientists were trying to comprehend the role of these molecules on the surface of cells in self-recognition and in the coordination of cellular activities in the individual. "These structures are, in fact," Dausset claimed, "the identity card of the entire organism."[37]

Before World War II, the biology of susceptibility and resistance to tumors had excited considerable interest. The British geneticist J. B. S. Haldane speculated that the gene products controlling the fate of tumor grafts—their acceptance or rejection—might resemble the antigens on the red blood cells that allowed blood grouping. He advised a promising graduate student, Peter A. Gorer, to try to work it out, and bestowed on him the inbred mice he had acquired on a recent visit to the Jackson Laboratory in Bar Harbor, Maine—no others being available in England.[38] Using a simple agglutination test for antigens, Gorer first identified four blood groups in the mice. When a mouse developed a tumor, he inoculated the others with its tissue and found that all who died possessed blood group antigen II, even on ordinary cells of the body. This suggested to Gorer that he might have located the genes governing susceptibility and resistance to transplants, but his research was largely ignored.[39] In 1946, Gorer took a visiting fellowship at the Jackson Laboratory, where he worked with George D. Snell, who also was inquiring into what he called the "histocompatibility" genes. A former student of Harvard biologist William E. Castle, Snell had been building up an impressive colony of inbred mice in a few crowded, unembellished rooms at Bar Harbor.[40] In the bucolic simplicity of Maine, Gorer set about tissue typing the mice Snell had bred to resist tumor strains. Snell and Gorer discovered that the genes coding for transplant immunity in the inbred mice were the same as those Gorer had identified in London, so the pair compromised and called the presumed genetic locus histocompatibility-II, or H2.[41] For the following decade, the H2 system kept expanding until it boasted more than thirty antigens. "For those who did not belong to the H2 fraternity," Leslie Brent complained, "listening to expert discussions on H2 tended to be like listening to people speaking a foreign language."[42]

While working in the ambulance service of the Free French Forces in North Africa, Jean Dausset became concerned about idiosyncratic reactions to the blood transfusions he performed. In Paris after the war, he dedicated himself to research on the immune responses of multiply transfused patients—and, somewhat more fitfully, to running a surrealist art gallery in St-Germain.[43] In 1952, he noticed that when blood serum from a transfused patient was placed on a glass slide and mixed with white blood cells, leukocytes, from someone else, the introduced leukocytes clumped. Evidently the serum of the multiply transfused patient contained antibodies against some antigen on the surface of the leukocytes; he named the antibodies MAC, for the initials of the leukocyte donors.[44] Over the following years, Dausset tried to make sense of this antigenic difference between individuals. In 1958, he studied an anemic patient who had received multiple transfusions from one donor: several weeks later, the patient was producing antibodies against a human leukocyte antigen (HLA) present in a proportion of the French population, an antigen later called HLA-A2.[45] In the 1960s, scientists confirmed that leukocyte surface antigens, unlike the familiar red blood cell antigens (the ABO system), were shared with most tissues in an individual body. As the decade wore on, more and more tissue antigens were announced, with each group of researchers asserting its own priority and unique nomenclature. As Brent noted, "the HLA arena was inhabited by some powerful personalities and, not surprisingly, there was a strong competitive edge to many publications and discussions."[46] After 1964, the leading contenders, longing for assent to their system, arranged annual international histocompatibility workshops to bring some order to the research enterprise and to standardize classification.[47] By the end of the decade everyone conceded that only one system of leukocyte antigens existed, and it was dubbed the major histocompatibility complex (MHC).

Dausset and others had understood at once the significance of human leukocyte antigens—the human equivalents of the mouse H2 system—for tissue transplantation. Thanks to their research, we know that virtually every human possesses a different combination of HLA genes. The more alike the tissue antigens of host and donor, the less likely a graft will incite a devastating immune reaction and therefore not be successful.[48] Accordingly, many investigators focused on refining and stabilizing tissue-typing methods in order to improve the odds of successful organ transplantation. Some immunologists, Dausset included, chose also to puzzle over the evolutionary significance of tissue antigens: since they had been conserved, they presumably served some purpose other than to impede transplantation and thus frustrate ambitious surgeons. With a surrealist eye for unexpected juxtaposition and variation, Dausset became obsessed with mapping patterns of HLA expression around the world. Geneticist Luca Cavalli-Sforza joined him in the hopeful effort to trace the global distribution of human polymorphism, or genome diversity.[49]

Others found clinical immunology more appealing. Experience with an asthmatic cough as a child made Baruj Benacerraf, a Venezuelan-American immunologist and part-time banker, curious about allergy and the mechanisms of hypersensitivity. After completing his medical degree at the University of Virginia, he took the advice of René Dubos and Jules Freund to seek out Elvin Kabat in New York, someone who could teach him about immunological disorders.[50] In the 1960s, he discovered that when he injected randomly bred guinea pigs with a foreign substance, almost 40 percent of them failed to mount an immune response. He suspected that genes might be controlling their immune responsiveness, a novel idea at the time. After moving to the National Institute of Allergy and Infectious Diseases in Bethesda, Maryland, Benacerraf acquired a stable of inbred guinea pigs, a legacy of the biologist Sewell Wright, which accelerated his efforts to map the supposed immune-response ("Ir") genes.[51] Meanwhile, in the Stanford Medical School, Hugh O. McDevitt, an Irish-American from Ohio, also was inquiring into genetic control of the immune response. In the late 1960s, he bought from the Jackson Laboratory some inbred mice that Snell had contrived to differ in only one genetic locus, at H2. To his astonishment, McDevitt found that their response to a standard antigen—a synthetic polypeptide—depended on this exiguous difference in the histocompatibility genes. The immune response genes appeared linked to, if not overlapping, the H2 complex. The improbable connection of genes for transplantation antigens with genes for the immune response was fascinating.[52] "The biological significance of this remarkable relation between these two highly polymorphic specificity systems," Benacerraf and McDevitt wrote in 1972, "is not understood"—but it could hardly be "fortuitous." Brought together by common genetic interests, they speculated on whether they had discovered "a large number of histocompatibility-linked genes that control the specific immune response."[53] But they admitted that the mechanism would remain an enigma until someone found the product of the immune response genes.[54]

Immunologists in the early 1970s felt the plot thickening. McDevitt linked up with his geneticist colleague Walter F. Bodmer to study the association of HLA with disorders of immunological reactivity. In 1971, they determined that people with the histocompatibility genes A8 (now B8) and W15 (now B15) were more likely to develop systemic lupus erythematosus compared to a control population with other tissue antigens.[55] The same year, Noel Rose revealed that some H2 genes predisposed mice to autoimmune thyroiditis.[56] Soon, Paul I. Terasaki in Los Angeles demonstrated an association of HLA and multiple sclerosis, and Ian Mackay in Melbourne announced the connection of HLA and autoimmune hepatitis.[57] As it turned out, histocompatibility genes correlate with susceptibility to many other diseases, especially other autoimmune diseases, including rheumatoid arthritis and type 1 diabetes.[58] This genetic relationship seems to offer at least a partial explanation for the propensity of certain families to suffer autoimmune disease, an unfortunate aggregation well known to clinicians. Yet, the association of tissue type and autoimmunity is not absolute. Not all cases of an autoimmune disease possess the indicted HLA configuration; and in only a few of those persons bearing a predisposing HLA arrangement does autoimmune disease ensue. Evidently, some additional exposure to unknown environmental factors—perhaps infection—is required to trigger disease. Moreover, it is likely that other genes contribute to autoimmune susceptibility. The striking association of tissue type and autoimmune disease nonetheless opened a new window onto the pathogenesis of these conditions, renewing an interest—languishing for most of the twentieth century—in what once was called diathesis, the constitutional predisposition to disease. "There is every reason to believe," McDevitt and Bodmer wrote wishfully in 1974, "that

HLA-linked specific immune-response genes will be shown to be important genetic factors predisposing to resistance and susceptibility to a variety of neoplastic [cancerous], autoimmune, and infectious diseases in man."[59]

The general immunological function of tissue (or histocompatibility) antigens was becoming clearer, even if aspects of their connection to autoimmune disease remained obscure. Thrown together in the Department of Microbiology at the Australian National University in Canberra in the early 1970s, Rolf Zinkernagel and Peter C. Doherty set about finding a reliable assay in mice of cytotoxic T cell activity, in order to investigate cell-mediated immunity to viruses, especially the lymphocytic choriomeningitis virus.[60] Having demonstrated potent cytotoxic T cell activity in virally infected mice, the young scientists decided to compare effects in different mouse stocks, each group displaying dissimilar tissue antigens. When they introduced T cells to various virus-infected cell cultures, cellular destruction occurred only when the T cells and the target cultured cells shared H2 antigens.[61] Injecting T cells sensitive to the virus into immunosuppressed, virus-infected mice, Zinkernagel and Doherty found the lymphocytes continued to multiply only if they shared on their surface the same H2 antigens as the host mouse. The assays indicated that the histocompatibility genes "restricted," or rendered possible, the cytotoxic T cell response: that is, the cytotoxic T cells killed if they were MHC compatible but not if they were MHC different. Zinkernagel and Doherty proposed that responding T cells must carry the same histocompatibility molecules as the cells—whether normal cells, macrophages, or dendritic cells—that were presenting foreign antigens, such as viruses, for the delectation of the lymphocytes. Thus, histocompatibility antigens promote and safeguard cell cooperation. To trigger an immune response, the foreign antigen has to be delivered to the cytotoxic lymphocytes along with the individual's own tissue antigens.[62] Another way of putting this is to say that cytotoxic T cells, in effect, are recognizing an altered self—or that the foreign is always recognized, almost in a solipsistic fashion, in the context of self. Therefore, it is really all about the self, and not, as Burnet argued, a matter of becoming aware of a non-self intruder etched in relief by a forbidden, or nonimmunogenic, self. At last, the immunological function of the tissue antigens was evident. Once thought of as transplantation antigens, they now were cast as part of a larger surveillance system. According to Doherty, "the key biological role of the H2 antigens was not to be 'seen' as 'foreign' on a grafted tissue, but to be the target for some self-monitoring, or immune surveillance, function concerned with eliminating abnormal cells within, say, a virus-infected person."[63] Or as Benacerraf wrote, "T cell immune responses are primarily responsible for monitoring self and non-self on cell surfaces."[64]

❋ ❋ ❋

At the end of the century, immunology still seemed fixed on the self, yet it was a far more dynamic, interactive, and positive self than Burnet could have imagined, an immunological identity still in formation, still learning, shaped by its continuing history.[65] "Antigen-sensitive cells with anti-self and anti-non-self specificity appear at a given moment and continue to be generated throughout the life of the animal," Melvin Cohn wrote from the Salk Institute in La Jolla, California. "This is a consequence of . . . the fact that the immune system must function in a constantly changing self and non-self world." In adult life, Cohn continued, the "immune system must periodically sample the self-environment and constantly re-correct its catalogue of specificities."[66] Embryonic processes and contingencies have loosened their grip on the immunological self. At the same time, scientists discern with greater clarity the cellular interactions involved in sustaining self-recognition, and its complex of molecular mechanisms. Ironically, immunological research constructed this version of human individuality through studies of mouse models and other veterinary analogues. Inbred mice and fowl became the principal sources of our knowledge of lymphocyte interactions and molecular configurations in the human immune system. Veterinary procedures allowed recognition of self in the human immune system to be translated into positive cellular and molecular terms.[67]

In the 1970s and 1980s, it became fashionable to assemble immunological networks.[68] "The immune system achieves a steady state," Niels K. Jerne wrote in 1974, "as its elements interact between themselves, and as some elements decay and new ones emerge." In Basel, Switzerland, Jerne summoned up an "idiotypic" network, a gathering of immunologically responsive parts of antibody molecules and lymphocytes. He observed the striking resemblance of the immune system and the nervous system, both of them dispersed and sensitive to an enormous variety of signals, though lymphocytes are more abundant and mobile than nerve cells. Both systems "learn from experience and build up a memory that is sustained by reinforcement and that is deposited in permanent network modifications, which cannot be transmitted to our offspring."[69] Both systems frame the self. According to Jerne, immunological discrimination between self and non-self depends partly on the deletion of self-reactive clones in the thymus and partly on the production of anti-idiotypic antibodies against autoantibodies and self-reactive T cell receptors. The immune network is complete; it can counter anything, even itself. Jerne believed that "the immune system (like the brain) reflects first ourselves, then produces a reflection of this reflection, and that subsequently it reflects the outside world: a hall of mirrors."[70] But Cohn spoke for many ordinary immunologists when he claimed that Jerne's idiotypic network was "largely smoke and mirrors" and "lacked logic and rationale."[71] He decried "the misdirected bandwagons engendered by idiotypic networks, suppressor circuitry and transcendental repertoires."[72] In 1980, Jerne gained the Nobel Prize in Physiology or Medicine, but his idiotyic network risked being consigned to the scrap heap at the end of the following decade.

Other immunological networks proved somewhat more durable. For Francisco J. Varela and Antonio Coutinho, all immune events are predicated on self-recognition; everything else is noise or nonsense. Like Jerne, they regarded the immune system as a network, not as isolated clones of lymphocytes. Varela viewed the immune system "as a closed network of interactions which self-determines its ongoing pattern of stability and its capacities of interaction with its environment."[73] In this network it is not so much a matter of differentiating self and non-self, or even ensuring self-tolerance, as invoking self to appraise hazard. "The presence of foreign materials in the organism," according to Varela and colleagues, "can only acquire immunological relevance by interaction with components of the immunological self. All immune events will be understood as *self-referential*, performed in reference to the self."[74] The immune system constantly monitors the self in order to recognize altered self or non-self. In the "cognitive domain" of the immune system, the recognition of altered self is not just molecular or structural, as it seemed to Zinkernagel and Doherty, but functional, the result of changes in relationships in the network—a perturbation of the "ecology of lymphocytes."[75] It is necessary, Varela claims, to "see how such a self can, at the same time, be a virtual point with no localized coordinates, and yet provide a mode of identity through which an interaction can happen."[76]

Israeli immunologist Irun R. Cohen elaborated on this self-referential, contextualist perspective. For him, a library of key self-antigens traces a sort of immunological homunculus, a physiological image of the self, encoded in the immune network. The immunological homunculus is a minimalist and stereotypical impression, oriented around self-antigens, controlled by the regulatory machinery of the immune system. In contrast, the clonal selection theory adumbrated a self that was "virtual and without substance," a subject defined by its ground, the foreign; accordingly, autoimmune disease occurred when the virtual self-antigen became a real self-antigen.[77] For Burnet, the self was a state of exception, of bare life; whereas for contextualists like Cohen, it is a disciplinary, governed condition—cultivated life.[78] Within Cohen's scheme, "the selfness or foreignness of an antigen depends on the interpretation given it by the immune system." Thus, "immunological disease results more often from a defect in interpretation than a defect in recognition."[79] If a self-antigen is presented in the context of infection, for example, it can appear non-self. Context and control are fundamental. According to Cohen, "the immune system must be sensitive to the state of the body and intimately responsive to it; autoimmunity, rather than shunned, is built into the system; a degree of autoimmunity must be physiological."[80] But in the wrong context, with adjuvants, it could become pathological.

For some immunologists, the network came to displace self altogether. At Yale, Charles A. Janeway, Jr., pointed out that not every foreign substance gives rise to an immune reaction; for example, silicone, bone fragments, and food usually are ignored. At the same time, the immune system generates antibodies against many ostensibly "self" antigens.[81] Janeway attempted to salvage the distinction by proposing that innate antigen-presenting cells, typically dendritic cells, recognize dangerous infectious non-self "stranger signals," thus reasserting a pragmatic division between self and other.[82] He failed to assuage Polly Matzinger's objection. "The immune system does not care about self and non-self," she declared in 1994. "Its primary driving force is the need to detect and protect against danger." In order to do so, the immune system "receives positive and negative communications from an extended network of other bodily tissues."[83] The obsession with the distinction between self and non-self had obscured the crucial question of what criteria the immune system uses in deciding to attack. According to Matzinger, an immune reaction required discharge from ordinary body tissues of danger signals, indicators of cell damage, to rouse and arm the antigen-presenting cells. The immune system is "not simply a collection of specialized cells that patrol the rest of the body, but an extended and intricately connected family of cell types involving almost every bodily tissue." Tolerance of one's body therefore is a "cooperative endeavor among lymphocytes, APCs [antigen-presenting cells], and other tissues."[84] While we all harbor autoreactive lymphocytes, these lurking cells demand evidence of damage—an alarm signal—before being spurred on to bring about autoimmune disease. "When healthy, tissues induce tolerance. When distressed, they stimulate immunity."[85]

Many immunologists found "danger theory" rather vapid and tame. To be sure, it possessed considerable appeal to those who believe the immune system ought to focus on potential harm, not fritter away its energy in discrimination of self. But Matzinger struggled to define danger and to avoid entrapment in circular logic. Some immune responses occur without tissue damage, as in the reaction to vaccines; and often tissue damage is the result, not the cause, of the immune response. Inasmuch as whatever activates an immune reaction can be designated dangerous, there is a risk of tautology. Proponents of the danger theory boast of their inclusion of antigen-presenting cells and ordinary tissues in the greater immune network, but there is nothing to stop contextualists, who are still invested in altered self, from encompassing these, too. Some immunologists disparage the danger theory as vague, "reductionist," and lacking "critical depth"; for them, self remains at least a "useful heuristic device."[86] According to Zinkernagel, "when used in a loose way, the word 'danger' gains the quality of a dogma that comprises a multiplicity of assumptions and catches the imagination."[87]

In a sense, the advocates of danger theory are trying to reconstruct a sharp divide between the normal and the pathological, to identify the abnormal molecular prompts to immunological responsiveness. They are reacting against the idea of immunity as a biographical project; they express discomfort with the notions that sensitivity to self and altered self are guiding immunological responsiveness and that pathological reactivity is simply a gradation of or deviation from normal self-referential activity. "I am a reaction to what I am," wrote French poet and philosopher Paul Valéry. "'What I am' is what appears to that which will be 'what I am.'"[88] Instead, danger theorists seek the molecular ontology of damage, the atomic form of an otherwise metaphysical nemesis. Thus, pathology is not monist exaggeration or alteration of normal physiology; rather, it is opposed to the normal, discontinuous with it. In danger theory, disturbance of function is subordinated to the lesion. Forestalling the self, disregarding reactivity and idiosyncrasy, danger models return us to the conventional—and reductive—microbiological view of disease.[89]

Foundering in philosophical deep water, ordinary immunologists have frequently expressed distaste for grandiose, even metaphysical, claims, and returned happily to the laboratory bench. "The Olympian heights that could be accessed by a Burnet or a Jerne," Peter Doherty observes, "look out on a vista that is currently so obscured by complexity and detail that broad generalizations elude us. Like traveling to some exotic destination that was once unique but now features a horrible combination of tacky sameness, maniacal traffic and air pollution, climbing that particular Mount Olympus isn't worth the effort."[90] Although still creditable, theories of self and non-self, like other ambitious biological and philosophical notions, might seem immaterial to the practical inquiries of most immunologists in the new century.

❋ ❋ ❋

Late-twentieth-century immunology could be arcane and strangely fractious, but on a practical level everything was coming together. In the 1990s, most immunologists agreed on the basics. In early life, the thymus generates T cells with an enormous range of variation in the receptors on their surface, while the bone marrow generates precursors of antibody-producing B cells. Those lymphocytes (whether in thymus or bone marrow) with unusually high sensitivity to self-antigens are eliminated, while lymphocytes possessing low affinity receptors receive a license to operate.[91] Normally, this low affinity for tissue antigens proves insufficient to activate helper and cytotoxic T lymphocytes without the stimulus of a foreign antigen. This means the thymic lymphocytes usually are both histocompatibility restricted and self-tolerant. In the periphery, a normal immune response begins when an offending antigen is consumed by cells, including macrophages and dendritic cells, and presented on the cell surface alongside tissue antigens. Such a combination proves irresistible for a matching lymphocyte. It activates specific clonal proliferation of T and B lymphocytes, generating a cytotoxic T cell response or arousing helper T cells to enhance B cell antibody production. The immune system thus displays a redundant repertoire of protective mechanisms but also the potential to go awry and cause damage. "Beneficial immune protection," Zinkernagel argues, "must therefore be balanced in evolution with potentially lethal damage by immune responses."[92]

During the twentieth century, autoimmunity has progressed from a prohibited occurrence, to an uncommon pathology, to a normal process—from never, to sometimes, to always.[93] Thymic lymphocytes are fashioned to recognize altered self-antigens, so the foreign is presented to them tied to molecules that mark the self. Even though autoimmunity is thus the physiological means of determining the immune system repertoire and ensuring its activation, rarely does it shade into autoimmune disease. To be sure, natural autoantibodies can be detected in the body, produced by rogue B cells—but regulatory T cells, disobliging helper T cells, and perhaps even other antibodies generally keep them in check. Failure of these control mechanisms, resulting in autoimmune disease, often correlates with the presence of certain tissue antigens that seem linked to supererogatory immunological potential of one sort or another, which in normal circumstances might be advantageous. And yet, while many might thus be predisposed to autoimmune disease, few succumb. Unlucky exposure to an infectious agent or environmental factor, damage to normal tissue, or the revealing of a previously sequestered self-antigen, probably is also necessary to trigger autoimmune disease. In some cases, a perceived homology between self-structures and microbial antigens, a sort of accidental mimicry, could disrupt tolerance, setting off a slow pathological cascade. A lot of things have to go wrong, a lot of control mechanisms fail, for autoimmune disease to become manifest—though once established, it seldom reverses.[94]

AFTERWORD

Becoming Autoimmune, or Being Not

❋ ❋ ❋

Looking back over the past hundred and fifty years, so much has changed in our knowledge of diseases like systemic lupus erythematosus, rheumatoid arthritis, multiple sclerosis, and type 1 diabetes mellitus—so much that it is tempting to consider them as novel entities or reinvented forms. In the nineteenth century, medical investigators conferred specific names, often heatedly contested, on what once were diffuse and vague complaints; and then, in the twentieth century, they profoundly altered conceptions of how such inchoate symptoms technically came into being and clustered together. As the ranks of sufferers swelled, disease specialists struggled to understand these conditions in contemporary biomedical terms, each scientific generation reframing them in its own style, with every conceptual refinement further separating medical theory from the illness experience. Once putatively degenerative disorders, or the result of hereditary diatheses, these diseases shifted biomedical shape, turning eventually into pathologies of immunity—what happens when our defensive system goes awry and attacks the body's own tissues, when a normal process shades into abnormality or error.

Yet, for much of the twentieth century, the majority of microbially minded scientists and physicians had regarded skeptically the notion of autoimmunity, the idea that the body might be sensitive or reactive to its own components, allergic to itself. At first the phenomenon looked like a laboratory artifact, dependent on intricately assembled experimental chimeras, arcane tests, and mysterious adjuvants. But the expansion of clinical research after World War II impressed upon laboratory workers the medical implications of their biological speculations. Several scientist-clinicians seized on autoimmunity as the key to understanding the cause of previously enigmatic chronic ailments. Gradually the idea gained adherents, eventually coming to seem conventional. Indeed, autoimmunity became the pathological test of the standard explanation of what was supposed to be the normal function of the immune system, the differentiation of self and non-self. This stabilization or validation of the autoimmune thought style came about through a transnational project within biomedicine, requiring the coordination of scattered immunology laboratories across the world and the development, during the Cold War, of international networks of dedicated scientists and clinicians, all ardently preaching the gospel of self-reactivity.

Even as autoimmunity emerged, matured, and propagated in the twentieth century—acquiring bit by bit a more molecular, reductionist tone—the concept retained some antique properties of medical reasoning. Indeed, one might view the apparent conceptual transformation as a twentieth-century recapitulation, even if drawn from a different perspective, of earlier theories of pathogenic fevers, where a physiological process like the maintenance of body temperature could go wrong and appear to cause illness. To a remarkable degree, the notion of autoimmune disease, with its emphasis on idiosyncrasy and individual variation, represents the survival of older, Hippocratic beliefs in disease as a biographical process. An overlay of novelty has obscured persisting characteristics. The delicate shading from the normal immune response into pathological autoimmunity expresses some alteration, at once subtle and ruthless, in the biological constitution of the individual, a change in recognition of self. As a truly personalized medicine, diagnosis and treatment of autoimmune disease are predicated on the individuality of the pathological process. The common use of corticosteroids, which affect general immune responsiveness, in a dosage titrated in part against the patient's experience, vividly demonstrates the persistence of an element of biographical sensibility even in twentieth-century biomedicine. But these older constitutional or personalized inferences now tangle, inextricably, with molecular mechanisms.[1] Although one may discern a residue of hereditary diathesis in contemporary autoimmune explanations, it manifests in gene frequencies and histocompatibility profiles. There is, thus, a continuing productive tension—perhaps not resolvable—between resilient concepts of biological individuality and ever more specific and determinate molecular causal mechanisms and therapeutics. Idiosyncrasy now is dressed in uncomfortable molecular garb.

❋ ❋ ❋

In becoming autoimmune, how have the diseases we have considered in this book changed? Once attributed solely to some hereditary predisposition, systemic lupus erythematosus now is marked by the presence of antibodies directed against cell nuclei, the consequence of dysfunctional B lymphocyte activation, assisted by overexcited helper T lymphocytes, all of this cellular commotion incited by genetic and environmental interactions. Complexes of nuclear antigen and autoantibody from trigger-happy B cells are laid down in a variety of tissues, inducing inflammation and leading to widespread organ damage. But the environmental event that precipitates SLE remains obscure; nor is it clear precisely which genetic alteration gives rise to the abnormalities of the B and T cells. Corticosteroids and other immune suppressants are still the mainstays of treatment, although more closely targeted therapies, based on enhanced knowledge of molecular mechanisms, show considerable promise. The new therapeutics generally tries either to use antibodies against B cells to inhibit their autoantibody production, or to find biochemical means to block immune cell stimulatory molecules.[2] Regardless of the impressive molecular elucidation, Flannery O'Connor would find few surprises in the current clinical management of systemic lupus.[3]

Similarly, rheumatoid arthritis has undergone the transformation from a diathesis or, as intermittently postulated, an infection to an autoimmune disease. The presence of rheumatoid factor, an autoantibody that causes inflammation in the joint lining, usually is a reliable marker, though its absence, intriguingly, does not exclude the diagnosis. Strongly associated with certain histocompatibility types, rheumatoid arthritis seems to require the continued activation of T cells, stimulating B cells to produce destructive autoantibodies, which escape effective regulation. The stimulus may be an infection or injury affecting a joint, but no one knows for sure. Suppression of inflammation remains the first-line treatment, and most patients will need methotrexate or some other anti-inflammatory agent to modulate the immune response. The new, "biologic" drugs targeting pro-inflammatory molecules—such as tumor necrosis factor—have proven remarkably effective, transfiguring the clinical management of rheumatoid arthritis and making its therapeutics at once more specific, more individual, and more profitable.[4]

For Barbellion, multiple sclerosis was a "deliberate, slow-moving malignity," perhaps the manifestation of bacteria in his spine or a symptom of constitutional degeneration.[5] Now, the demyelinating disease is regarded as the result of an autoimmune response in a genetically susceptible individual. Without any definitive immunological test, the diagnosis of multiple sclerosis, a highly variable condition, depends on clinical history, neurological examination, and findings of magnetic resonance imaging. The uneven geographical distribution of the disease implies some combination of genetic and environmental influences, perhaps related to vitamin D metabolism, perhaps involving an initial infection. Overstimulated, autoreactive T cells, migrating across the blood-brain barrier into the central nervous system, are the principal cause of continuing damage to the myelin coating of nerves. Treatment of multiple sclerosis largely remains supportive, addressing the various neurological problems, although immune modulating agents, including interferon and T-cell inhibitors, recently have been used to alter the course of the disease. As pathology takes hold and the nerve cells deteriorate, even the modest benefits of immune suppression rapidly diminish.[6]

During the twentieth century, type 1 diabetes, like so many autoimmune diseases, has become far more prevalent—for reasons still obscure. A condition commonly associated with onset early in life, sometimes in infancy, autoimmune diabetes can suddenly afflict previously healthy adults, too. Before the development of symptoms, antibodies attack and destroy the insulin-producing islet cells of the pancreas; by the time symptoms appear, most of the damage has been done. It is unclear what causes the breakdown of normal tolerance of the islet cells and the activation of T cells that stimulates antibody production. Again, there is a strong connection with tissue type (MHC), suggesting hereditary predisposition. There has been intense argument about the role of viruses in triggering the autoimmune reaction. As the immunopathology is often exhausted by the time diagnosis is made, the emphasis is on insulin replacement and treatment of secondary problems, not on immune suppression—though stimulation of regulatory T cells in the early stages of disease has been tried.[7]

The unraveling of the molecular and cellular mechanisms of autoimmune disease, the unscrambling of its pathogenesis, contrasts with continuing speculation on what could trigger the problem; these research achievements sit uneasily alongside enduring perplexity concerning etiology. We have a good idea of how autoimmune disease develops, and we can even intervene on occasion to retard or modify it, but we still know little about what provokes the process. In the face of this uncertainty, there is generally a default to old axioms of hereditary susceptibility and environmental influence, perhaps involving infection and tissue damage. And so, again, amidst conceptual transformation and elaboration, we find a lasting medical dialectic; only now, heredity has gained molecular complexity through the interplay of genetics and epigenetics and the environment has become more animated, including microbes, whether alien or resident within us.[8] It seems that something stimulates a defensive reaction that refuses to stop; the immune system mistakes normal tissue for whatever needs to be eliminated, maybe perversely misrecognizing the self as something foreign. The misguided overreaction then somehow escapes normal regulation. Evidently, there are many possible causes or incitants of autoimmune disease, but as yet, none has been identified with precision and authority. The etiological framework is sturdy enough, but it is empty. Those who suffer have to find their own answers to the existential questions Why me? and Why now?

❋ ❋ ❋

One of the themes in our recounting of the emergence of autoimmunity is the recurrent passage of metaphors back and forth between philosophy and science, the intermingling of concept and model between domains that conventionally are held separate. For most of the twentieth century, metaphors like "immunity" and "self" were prone to drifting from philosophy and social theory into the laboratory and clinic. "Immunology has always seemed to me," reflected F. Macfarlane Burnet, "more a problem in philosophy than a practical science."[9] He agreed with Alfred North Whitehead, who warned that "if science is not to degenerate into a medley of *ad hoc* hypotheses, it must become philosophical."[10] In Burnet's wake, however, scholars have taken the opposite path, making their philosophy and social theory more immunological. As the science of the self, immunology offered a rich vocabulary and an attractive conceptual framework for an elevated discussion of human identity. The concept of immunity—returning, in a sense, from whence it came—gained even more power as an organizing principle, and abundant source of metaphor, for social theory at the end of the twentieth century. According to critical scholars like Donna Haraway, "the immune system is an elaborate icon for principal systems of symbolic and material 'difference' in late capitalism." It draws a map "to guide recognition and misrecognition of self and other in the dialectics of Western biopolitics."[11] For anthropologist Emily Martin, "the science of immunology is helping to render a kind of aesthetic or architecture for our bodies that captures some of the essential features of flexible accumulation [in capitalism]."[12] In his last years, philosopher Jacques Derrida "granted to this autoimmune schema a range without limits."[13] Derrida especially appreciated the "undecidability" and internal contradiction of autoimmune processes, the way in which they naturally deconstructed supposedly sovereign bodies and commanding identities. "We feel ourselves authorized to speak of a sort of general logic of autoimmunization," wrote Derrida in 1996. "It seems indispensable to us today for thinking the relations between faith and knowledge, religion and science, as well as the duplicity of sources in general."[14] In the new millennium, it appeared, we were all constitutively autoimmune, both bodily and socially. Certainly, many Western intellectuals were finding it increasingly hard to consider life and death without invoking the immunological self and autoimmunity. Their immunological enthusiasm perhaps exemplifies Ludwig Wittgenstein's observation that "work in philosophy is . . . a kind of work on oneself"; that the treatment of a philosophical problem is "like the treatment of an illness."[15]

The scientific idea of autoimmunity had emerged from the early-twentieth-century concerns of Viennese and Parisian intellectuals with sensitivity and reactivity, with allergy and anaphylaxis, with idiosyncrasy and biography. Between the wars, autoimmunity became entangled with organicist thinking and process philosophy, with biological speculations on individuality and pattern. During the Cold War, autoimmunity bonded with cybernetics and theories of command and control. While many scientists reveled, as Burnet did, in the mixing of metaphor and model between biology, philosophy, and local culture, others have found such promiscuous exchange disturbing and disorienting. Historian Ilana Löwy counts the notion of self as a useful "heuristic tool of imprecise definitions"; yet philosopher Alfred I. Tauber worries that *self* is a word borrowed from philosophy "to approximate a language that is inadequate to the task," a seductive, opportunistic metaphor.[16] Even Derrida cautions that autoimmunity "might look like a generalization, without any external limit, of a biological or physiological model"; but then he rejects any purification of his terms and goes on to deploy the idea with abandon.[17] These admonitions against metaphoric borrowing, albeit often qualified and equivocal, prompt us to ask what work immunology might be doing in philosophy. How should this scientific knowledge inform or structure thinking about the way we live now? Does how we know or experience autoimmunity make a difference?

In the 1980s, many humanities scholars felt dissatisfied with older preoccupations with the disembodied self. They were searching the horizon for some fully corporeal alternative; immunology came to their rescue. According to Burnet, immunity "seems to be part of the process by which the structural and functional integrity of every complex organism is maintained"—it bodily constitutes the self.[18] But Burnet's clonal selection theory appeared too static, negative, and deterministic to elicit lasting enthusiasm among poststructuralists. The networked immune system proposed by Niels Jerne caused more excitation among anthropologists and social theorists looking for a new—and perhaps more flexible—way to explain identity. Conventionally, philosophers had tended to associate personal identity with the mind: long ago, John Locke had emptied the self from the body, separating it from later thoughts of immunity, which ensures corporeal integrity.[19] In the twentieth century, the more materialist philosophers came to locate human individuality in the brain, the mysterious province of neuroscientists. Consigned now to the brain, the self returned to just one part of the body. Historians of the neurosciences frequently observed the "reduction of self to consciousness as a function of soul or brain."[20] But fresh speculations on the immune system toward the end of the century postulated a fully embodied self, a body politic—a biopolitical individuality—available to intellectuals critically concerned with the human condition.[21] "For me, practical metaphysics has to be translated into the language of general immunology," claimed German philosopher Peter Sloterdijk, "because human beings, due to their openness to the world, are extremely vulnerable—from a biological level, to the juridical and social levels, to the symbolic and ritual levels. . . . The task of building convincing immune systems is so broad and all-encompassing that there is no space left for nostalgic longings."[22]

"The immune system," declared feminist scholar Donna Haraway in 1989, "is both an iconic myth object in high-technology culture and a subject of research and clinical practice of the first importance." Immunology texts and research reports published in the 1980s demonstrated that the immune system was "unambiguously a postmodern object," a coded system—a coding "trickster" even—for recognition of self and other.[23] Haraway pondered how distinctions between the normal and the pathological worked when the biological body was defined as a "coded text, organized as an engineering system, ordered by a fluid and dispersed command-control network." The hierarchical, organic, modern body became a matter of codes, dispersal, and networking—a semiotic system. "Disease is a subspecies of information malfunction or communications pathology," Haraway observed; "disease is a process of misrecognition or transgression of the boundaries of a strategic assemblage called self. . . . Individuality is a strategic defense problem."[24] Once simply a modernist immunological battlefield, the "hierarchical body of old has given way to a network-body of truly amazing complexity and specificity."[25] For Haraway and her followers, immunology after Jerne was creating an exciting postmodern body, reanimating human identity, writing a new constitution.

Exploring the images dominating end-of-century popular and scientific discussions of the immune system, Emily Martin tried to understand what sort of social world immunologists were conjuring. She believed that immunology excelled at rendering "natural" certain social arrangements and cultural assumptions. In popular accounts, the metaphor of warfare against an external enemy still prevailed; the body resembled a police state, protecting against foreign intruders. The boundary between self and other was rigid and absolute. These images of immunity make "violent destruction seem ordinary and part of the necessity of daily life."[26] In contrast, immunologists increasingly were inclined to depict the body as a "whole, interconnected system complete unto itself," as a "homeostatic, self-regulating system." Martin saw older, militaristic models of the body, "organized around nationhood, warfare, gender, race, and class," contending with a new immunological body "organized as a global system with no internal boundaries and characterized by rapid flexible response." It was, for her, a new body transformed for "late capitalism."[27] Martin was convinced that immunologists did not "ignore the world outside the lab in devising their models of the body." The cultural anthropologist recorded how the language of immunity, after Jerne, "crashed into contemporary descriptions of the economy of the late twentieth century with a focus on flexible specialization, flexible production, and flexible rapid response to an ever-changing market with specific, tailor-made products."[28]

The cultural anthropology of immunology has continued to prove alluring. According to A. David Napier, "immunological ideas now provide the primary conceptual framework in which human relations take place in the contemporary world."[29] He noted that Burnet's clonal selection theory had been predicated on a "culture-bound" assumption of a wholly autonomous "self" but that this "prior and persistent" identity now was superseded in immunology.[30] Immunologists had come to recognize that "self is made up and defined by potentially dangerous encounters at one's boundaries." In fact, the immune system appears to be exploring otherness as much as defending self; it is, indeed, involved in a "*creative attempt to engage difference*," not eliminate it. Antibodies, according to Napier, "function as *'self' search engines—search engines for the information (harmful or helpful) that sits latently in viruses*."[31] Like many humanities scholars, Napier is convinced that immunology "is perhaps better positioned than any other domain of modern science . . . to help us rethink notions of the self that have dominated Western philosophy at least since the Enlightenment." As a humanist, he thinks it timely to "reconsider immunology's contribution to the metaphysics of identity."[32]

Known for his practice of "deconstructing" or complicating the binary logic of philosophy and adept at revealing how the excluded other comes to haunt even supposedly pure or "unscathed" objects, Jacques Derrida discovered autoimmunity in the early 1990s. In

Specters of Marx, the roguish philosopher observed: "To protect its life, to constitute itself as unique living ego, to relate, as the same, to itself, it [the ego] is necessarily led to welcome the other within . . . it must therefore take the immune defenses apparently meant for the non-ego, the enemy, the opposite, the adversary and direct them at once *for itself and against itself*." According to Derrida, "the living ego is autoimmune [*le moi vivant est auto-immune*]."[33] He meant that an internal process was compromising the integrity of the person, dishonoring sovereign identity, overturning a power, at the same time as it opened up possible options for the future, for individual transformation. Before long, Derrida became obsessed with the logic of autoimmunity, the vision of a specter haunting the self. He perceived the need to let the ghosts speak, and he wondered "how to give them back speech, even if it is in oneself, in the other, in the other in oneself"—how, that is, to recognize and express an autoimmune logic.[34] Initially puzzled, his admirers eventually assimilated the immunological trope. Some came to regard the figure of autoimmunity as "an image with considerable surplus value, one whose immediate applicability is startling and that continues to resonate." They asked, "Is it the case that deconstruction itself is a species of autoimmunity?"[35] One of Derrida's English translators claimed that autoimmunity names "a process that is inevitably and irreducibly at work more or less everywhere, at the heart of every sovereign identity."[36]

At times, Derrida's understanding of autoimmune pathology was eccentric. "As for the process of autoimmunization, which interests us particularly here," he wrote in 1996, "it consists for a living organism, as is well known and in short, of protecting itself against its self-protection by destroying its own immune system."[37] Perhaps he was thinking of the transitory autoimmune explanation of acquired immunodeficiency syndrome (AIDS), which had proposed, wrongly as it turned out, that some immunological process was destroying the body's T lymphocytes.[38] In any case, the philosopher proved eager to extend the domain of immunological critique: "As the phenomenon of these antibodies is extended to a broader zone of pathology and as one resorts increasingly to the positive virtues of immuno-depressants destined to limit the mechanism of rejection and to facilitate the tolerance of certain organ transplants, we feel ourselves authorized to speak of a general logic of autoimmunization." "Nothing in *common*," he wrote, "nothing immune, safe and sound, *heilig* and holy, nothing unscathed in the most autonomous living present without a risk of autoimmunity."[39] For Derrida, autoimmunity could be a positive force, inasmuch as the Freudian death drive could be affirmative.[40] "Self-contesting attestation keeps the autoimmune community alive, which is to say open to something other and more than itself: the other, the future, death, freedom, the coming or love of the other."[41] In other words, autoimmunity stimulates us to rethink life and death.

Immunology also grips the Italian philosopher Roberto Esposito. "The fact that some of the most important contemporary authors, working independently from one another and following different paths of thought," he said in 2006, "came to work on the category of immunization signals just how important the category is today. . . . Today a philosophy that is capable of thinking its own moment [*tempo*] cannot avoid engaging with the topic of immunization." Indeed, for Esposito, the apprehension of immunity "invents modernity as a complex of categories able to solve the problem of safeguarding life."[42] In the past, immunology had been obsessed with violent defense against the foreign, the self being "modeled as a spatial entity protected by strict genetic boundaries," but now "the body is understood as a functioning construct that is open to continuous exchange with its environment." His understanding of Jerne and the more cognitively oriented immunologists leads Esposito to argue that "the immune system must be interpreted as an internal resonance chamber, like the diaphragm through which difference, as such, engages and traverses us."[43] The philosopher believes that one must choose to focus on "either the self-destructive revolt of immunity against itself or an opening to its converse, community"—on autoimmunity or tolerance.[44] It is through the self that one recognizes the other: "nothing is more inherently dedicated to communication than the immune system," writes Esposito.[45]

Often these philosophers and social theorists too easily associate Burnet and the Cold War with concepts of the static, predetermined genetic self and with militaristic metaphors, imagining a body in violent opposition to the other, constantly defending against the foreign. Yet Burnet, a meretricious straw man, persistently advocated the study of tolerance, including the investigation of peripheral, homeostatic mechanisms of exempting self—even if, on occasion, he defaulted to more bellicose language. Moreover, the Cold War long ago introduced cybernetics, theories of control and communication, to the study of self. Autoimmunity as a concept, disturbing assumptions of an inert self, deploying a fifth column on the terrain of the body, was gaining traction even in the 1950s. Philosophers, however, are inclined to attribute dynamic concepts of a communicative self, a self in the making, to the immunological network theories of the 1970s and later. Their history may be unreliable, their findings belated, but one has to admire the fervor that infuses their proclamation of the significance of immunity for our discernment of self and other, for our appreciation of security and danger, for the understanding of life and its contrary. In the twenty-first century, immunology—autoimmunity especially—seems applicable everywhere.

❋ ❋ ❋

"The most disturbing yet revealing instance of the body's betrayal of itself is autoimmune disease," writes American theologian Mark C. Taylor, reflecting on his own autoimmune diabetes mellitus.[46] The experience of disease led him to question his sense of self:

> The study of autoimmune disease implies that the classical scheme of creation-fall-redemption is deeply flawed. There is mounting evidence that autoimmunity is our "primal" condition and so-called health is secondary to, and dependent upon, the suppression of an autoimmune response. If this is so, then the self is never at one with others, its world, or even itself. Disease, in other words, is not "abnormal" but is always already in our midst.[47]

Certainly for Taylor, autoimmunity became the condition of life itself. Once a friend of Derrida, he has learned "to linger with a negative."[48] "If our initial relation to ourselves is autoimmunity," Taylor writes, "our body is not originally an integrated whole governed by the principle of inner teleology but is inherently torn, rent, sundered, and fragmented. The body is *always* betraying itself." For Taylor, it is truly a matter, a combined materiality, of life and death. "Betrayal is unavoidable, cure impossible. Disease is neither a mode of being nor of nonbeing but is a way of being not without not being. The dilemma, the abiding dilemma to which we are forever destined, is to live not."[49]

"I have been reading, thinking, and living the paradoxical logic of the autoimmune via the frame of a personal illness for some time now," writes Alice Andrews, an English cultural studies scholar suffering from systemic lupus erythematosus.[50] The words of Freud and Derrida shape her perceptions of the illness and shift her grasp on self. For Andrews, "the autoimmune names a vulnerability and risk that persists on both sides of a decision: to defend life and risk death?" She imagines her lupus as a "repetitious death-drive perhaps of an organism returning itself to the inanimate."[51] And yet:

> A "self-indulgent" concern for and return to the self is effected, perhaps, by the insistence of the body's symptoms that return in repetitive "flares," always unexpectedly, always painfully, always urgently, to interrupt my chain of thought: unexpected pains in my legs and joints immobilize me, a low-grade fever makes me tremble, and pleurisy catches my breath, I fidget as my skin itches and blisters, and my thoughts struggle through the headache to be drawn back, again, always, towards the terrible experience of discomfort.[52]

As Andrews explains, "lupus is an *auto*immune disease, and it admits a certain rogue-like wolfishness into the functioning of the organism." Using immuno-suppressants, a "counter-attack" is launched against her immunological "misfiring," although the treatment also can be dangerous.[53] Derrida's words, Andrews tells us, continue to echo: "there is no absolutely reliable prophylaxis against the autoimmune."[54] It is another iteration of Taylor's "abiding dilemma."

Both Taylor and Andrews are exceptionally sensitive to immunological philosophy, yet most people who suffer from autoimmune disease so far have displayed little interest in philosophical inquiries into self, or even in the scientific theories that fasten onto the concept. For most of them, as they try to get on with life, the overly complicated inner world of autoimmunity probably passes by unnoticed among the bustling medical crowd. For a few, to be sure, their diseases prompt speculation on self—though rarely is it especially philosophical or scientific in tone. "My blood plasma had filled with poison made by my immune system," writes Sarah Manguso. "My immune system was trying to destroy my nervous system. It was a misperception that caused me a lot of trouble."[55] In her early twenties, she acquired a chronic version of Guillain-Barré syndrome, when antibodies attacked the myelin sheaths of her peripheral nerves. "What came first," she wonders, "the suicidal depression or the suicidal autoimmune disease?"[56] After years of immuno-suppression, Manguso's disease eventually scuffled into remission. "The only thing I'd done in my life," she writes, "was recovering from a disease. My self-image had been highly susceptible to that event. It constituted most of my identity." But this "self" was existential, more than philosophical or technical. "Those who claim to write about something larger and more significant than the self," she cautions, "sometimes fail to comprehend the dimensions of a self."[57]

According to Australian historian Inga Clendinnen, "to feel the mind disintegrate and to fear the disintegration of the self, is to suffer an existential crisis, not a medical one."[58] Experiencing fatigue, headaches, nausea, dizziness, and stomach pains in her mid-fifties, Clendinnen received the surprising diagnosis of autoimmune hepatitis. At first, the ailment seemed "a new and stylish disease, rare, and dangerous." For a few months, "this disease was a kind of distinction," she writes. "This disease, I privately thought, had class."[59] But despite immuno-suppressive drugs her condition deteriorated. The writer became an invalid, teetering on the edge of liver failure, barely living "behind the invisible cordon of the chronically ill." "I had been dealt a disease which held the possibility of a dramatic denouement," she recalls. "My decline had been slowed by drugs. Beyond that artificial extension glimmered the prospect of the transplant operation."[60] During her illness, Clendinnen "used writing to cling to the shreds of the self," as she puts it. Then her long-awaited liver transplant was successful, though it sentenced her to a lifetime of further immuno-suppression. As she recovers, she finds that "both the energy and the desire to look further into the self are dead. This stuck together 'I' is tired of introspection, that interminable novel of the invention of the self."[61] She has little interest left in science, even less in philosophy. She concludes:

> So that is what I have been doing all this time—by courtesy of a physiological malfunction, taking a journey out, beyond and around myself, and into interior territories previously closed to me. At the end of it, battered, probably wiser, certainly wearier and, oddly, happier, I have returned to where I began: to history, with a deepened sense of what peculiar creatures we are, you and I, making our marks on paper, puzzling over the past and the present doings of our species, pursuing our peculiar passion for talking with strangers.[62]

And that is what we, too, have been doing all this time.

ACKNOWLEDGMENTS

The origin of this collaboration might be traced back to the early 1980s, when Warwick Anderson was a medical student and then an intern on ward 3E of the Royal Melbourne Hospital, the Clinical Research Unit of the Walter and Eliza Hall Institute, which Ian Mackay directed. We have remained in conversation about the history of investigations of the disease kuru and, more recently, the history of autoimmunity. As a result, this book is the culmination of more than thirty years of intellectual engagement between a clinician-historian and a historian-clinician.

We are particularly grateful to Charles Rosenberg and Jacqueline Wehmueller for encouraging us to write this book and for their careful guidance throughout the process. Two anonymous reviewers for Johns Hopkins University Press gave helpful suggestions for revision and clarification. We are indebted to those who generously read and commented on the whole book manuscript: Daniela Helbig, Cecily Hunter, Gustav Nossal, Maureen O'Malley, Hans Pols, Senga Whittingham, and Sybil Williams. Historians of science at Princeton University, led by Angela Creager, Michael Gordin, Katja Guenther, Erika Milam, and Keith Wailoo, kindly read the book in its penultimate version, offering useful and timely advice. Others have read and commented on sections of the book: they include Isabelle Baszanger, Tatiana Buklijas, Stacey Carter, Robert Clancy, Adele Clarke, Deborah Coen, the late Barbara Heslop, Ilana Löwy, Nicolas Rasmussen, Dale Smith, Mark Veitch, and Peter Winterton. Additionally, Melinda Cooper, Ian Kerridge, and Catherine Waldby gave encouragement and support at the University of Sydney. Spirited discussions with Thomas Pradeu and Fred Tauber helped to sharpen analysis in Chapter 4 and the Afterword. Through interviews and informal conversation, Peter

Doherty, Hal Holman, Jacques Miller, Gus Nossal, Ivan Roitt, and Noel Rose shaped our historical inquiries. Laura Lindgren found the cover image. This project thus became a more extensive and energetic collaboration than the title page indicates.

We could not have completed *Intolerant Bodies* without a group of extraordinarily helpful and dedicated research assistants: Edmund McMahon, Cecily Hunter, James Dunk, and David Robertson. Cecily did the bibliography, and Jamie compiled the index. At Monash, Elaine Pearson gave expert secretarial assistance to Ian Mackay.

Not a page of this book was written without thought of patients we have seen with autoimmune and other diseases. In particular, Ian Mackay could draw on his unrivaled clinical experience to relate our complex story to those most viscerally affected by it. We hope that those with autoimmune diseases will appreciate our efforts.

A small part of Chapter 2 appeared in a different form as Ian R. Mackay and Warwick Anderson, "What's in a Name? Experimental Encephalomyelitis: 'Allergic' or 'Autoimmune,'" *Journal of Neuroimmunology* 223 (2010): 1–4. An earlier version of Chapter 4 was published as Warwick Anderson and Ian R. Mackay, "Fashioning the Immunological Self: The Biological Individuality of F. Macfarlane Burnet," in the *Journal of the History of Biology* 47 (2014): 147–75. We thank the editors and reviewers of these journals for their constructive criticism.

The Australian Research Council generously funded our research through the award of a Discovery Project Grant, 2012–14 (DP 120100861). The University of Sydney and Monash University were wonderfully supportive environments for Warwick Anderson and Ian Mackay, respectively. Stephen Garton and Duncan Ivison helped provide for Warwick Anderson at Sydney, while Ed Byrne and Christina Mitchell looked after Ian Mackay at Monash. Warwick Anderson also thanks Hans Pols for providing a long-running writing retreat at Dangar Island, and Marcia and John Davies for hospitality in Melbourne. Ian Mackay is deeply grateful to Patricia Mackay for her resolute and affectionate tolerance throughout the writing of this book.

Unless otherwise stated, Warwick Anderson is responsible for translation from French, and Ian Mackay for translation from German.

We are pleased to dedicate this book to two great teachers and friends. Without the enthusiastic support of Barbara Gutmann Rosenkrantz, Warwick Anderson might never have dared venture into the history of immunology. Ian Mackay wishes to acknowledge the inspiration and guidance provided by his mentor, the inaugural director of the CRU, Ian Jeffreys Wood. As we were writing *Intolerant Bodies*, we kept turning to them as our ideal readers and interlocutors.

NOTES

Introduction: Thinking Autoimmunity

1. Heller and Vogel, *No Laughing Matter*, 19. Heller was 58 at the time.

2. Guillain-Barré syndrome was recognized as autoimmune in a pioneering early-1960s survey: Mackay and Burnet, *Autoimmune Diseases*.

3. Heller and Vogel, *No Laughing Matter*, 21.

4. Since there are more than eighty autoimmune diseases, a few of them being relatively slight and unreported, the epidemiology of autoimmunity is incomplete. A study of the prevalence in the United States of twenty-four common autoimmune diseases estimated (in the late 1990s) that each year more than 230,000 Americans develop one of these conditions: see Jacobson, Gange, Rose, and Graham, "Epidemiology and estimated population burden of selected autoimmune diseases." The prevalence of autoimmune disease probably is much the same elsewhere.

5. It is necessary to point out that the acquired immunodeficiency syndrome (AIDS) is not an autoimmune disease. Until the discovery of the human immunodeficiency virus (HIV), which infects some immunological cells, a few scientists had speculated that there might be an autoimmune reaction against parts of the immune system. AIDS is a disease of immunological deficiency, while autoimmune diseases are problems of overreactivity.

6. *Autoimmune* became an adjective in 1951: see Young, Miller, and Christian, "Clinical and laboratory observations." (*Autoallergic* was a common description in the 1940s—see Chapter 3.) For the first use of *autoimmunity*, see Anon., "The immunology of thyroid disease." Just a few months later, the same journal insisted on putting quotation marks around the word *autoimmune:* Mackay, Larkin, and Burnet, "Failure of 'autoimmune' antibody." Yet the following year, it confidently put the term in a heading: Anon., "Autoimmunity in thyroid disease."

7. For general studies of the history of immunology, see Silverstein, *A History of Immunology*; and Moulin, *Le dernier langage de la médecine.*

8. Temkin, "The scientific approach to disease"; Rosenberg, "The therapeutic revolution"; and Canguilhem, *The Normal and the Pathological.*

9. On the closely related history of allergy, see Jackson, *Allergy.*

10. See Zinsser, *Rats, Lice and History.*

11. On writing the history of the present, see Foucault, *Discipline and Punish.* We emphasize that this book is not a survey of contemporary scientific knowledge

about autoimmune disease. For that, the reader should consult Rose and Mackay, eds., *The Autoimmune Diseases*, 5th ed.

12. On historical epistemology, see Fleck, *Genesis and Development of a Scientific Fact*; Foucault, "Introduction"; Koselleck, *The Practice of Conceptual History*; and Rheinberger, *On Historicizing Epistemology.*

13. This accords with the suggestions of Anderson, Jackson, and Rosenkrantz, "Toward an unnatural history of immunology."

14. Ian R. Mackay has worked as a clinical immunologist since the mid-1950s; Warwick Anderson was his intern in 1984.

15. Manguso, *The Two Kinds of Decay*, 9.

16. Manguso, *Two Kinds of Decay*, 13, 131, 175–76.

Chapter 1. Physiology with Obstacles

1. Poe, "For Annie," 457; and Dickens, *Little Dorrit*, ii.

2. Ward observes "fever's ubiquitous place in novels, poetry, plays, and printed ephemera as well as medical writings" (*Desire and Disorder*, 15).

3. Dickens, *Great Expectations*, 343.

4. Dickens, *Bleak House*, 35. See also Gurney, "Disease as device"; and Benton, " 'And dying thus around us every day.' "

5. Dickens, *The Personal History of David Copperfield*, 51.

6. Bailin, *The Sickroom in Victorian Fiction*, 5. Perhaps the best example of this is Martineau, *Life in the Sickroom.* See also Stephens, *Notes from Sick Rooms.* For more on literary representation of fever, see Rothfield, *Vital Signs;* Wood, *Passion and Pathology;* Platts, "Some medical syndromes"; and Christensen, *Nineteenth-Century Narratives of Contagion.*

7. See the sufferers' accounts of autoimmune disease in Chapter 5.

8. Wilson, "Fevers and science."

9. Bronte, *Wuthering Heights;* Flaubert, *Madame Bovary;* Dostoyevsky, *The Brothers Karamazov.* See also Peterson, "Brain fever in nineteenth-century literature."

10. Bynum, "Cullen and the study of fevers"; and Pelling, *Cholera, Fever and English Medicine.* Cullen was adapting the theories of Georg Ernst Stahl and Friedrich Hoffman.

11. Bynum, *Science and the Practice of Medicine.*

12. Fordyce, *Five Dissertations on Fever*, 48, 5, 291.

13. Clutterbuck, "Remarks on Dr. Elliotson's clinical lecture on fever," 275. See also Clutterbuck, *An Enquiry into the Seat and Nature of Fever*, and *An Essay on Pyrexia and Symptomatic Fever.*

14. Thomson, *Lectures on Inflammation.* See also Brown, *Medical Essays on Fever.* The emphasis on the blood harks back to Herman Boerhaave.

15. Andral, "On the physical alterations of the blood," 419. See also Searle, "On the nature of inflammation."

16. Ackerknecht, "Broussais, or a forgotten medical revolution," and *Medicine at the Paris Hospital.*

17. Canguilhem, *The Normal and the Pathological.* See also Keating, "Georges Canguilhem's *The Normal and the Pathological.*"

18. Jones, "General considerations respecting fever," 644. See also his "Pathological and therapeutical considerations."

19. Corrigan, *Lectures on the Nature and Treatment of Fever,* 5, 34.

20. Ackerknecht, "Diathesis." See also Rosenberg, "The bitter fruit"; López-Beltrán, "Forging heredity"; and Waller, "'The illusion of an explanation.'" As Ackerknecht put it, "heredity has always been the facile way of explaining the inexplicable" (*Medicine at the Paris Hospital,* 63).

21. Gallwey, "Unhealthy inflammations," 307, 308.

22. Moxon, "Considerations bearing on our present knowledge," 931.

23. Tweedie, "Lectures on fevers. X: Mortality of continued fever," 488, 489. See also his *Clinical Illustrations of Fever;* and Budd, *On the Causes of Fevers.*

24. Travers, *The Physiology of Inflammation,* 23.

25. Liebig, *Animal Chemistry.* See also Pelling, *Cholera, Fever, and English Medicine.*

26. Maccalister, "Gulstonian lectures on the nature of fever," 507, 511. The emphasis on heat favors systemic, constitutional explanations; only with the development of clinical thermometry was it possible to differentiate temperature patterns in continued fevers and to see separate diseases. Carl August Wunderlich introduced the thermometer into his clinic in Leipzig in 1851, but its impact was delayed. See Wunderlich, *On the Temperature in Disease;* Stevenson, "Exemplary disease: the typhoid pattern"; and Hess, "Standardizing body temperature."

27. Moore, "On the production of heat in fever," 258.

28. Hegel, *Philosophy of Nature,* 434, 428. See Engelhardt, "Hegel's philosophical understanding of illness."

29. Hegel, *Philosophy of Nature,* 435.

30. Hegel, *Philosophy of Nature,* 435, 434. Other thinkers found the metaphor appealing. Goethe "had heard so much of cannon fever and wanted to know what kind of thing it was" (*Campaign in France in the Year 1792,* 728). Nietzsche believed that "we are all suffering from a consuming fever of history" ("On the uses and disadvantages of history for life," 60).

31. Ackerknecht, "Diathesis," 325.

32. Virchow, "Standpoints in scientific medicine," 26, and "Cellular pathology," 81.

33. Canguilhem argues that denial of disease ontology is "the deeper refusal to confirm evil" (*The Normal and the Pathological,* 104).

34. But, as we shall see, ontology, in the form of specific clinical classification, keeps reasserting itself. As Temkin remarked, "it is probably neither possible nor

advisable to renounce ontology completely" ("The scientific approach to disease," 450).

35. Virchow, "Cellular pathology," 87, and "One hundred years of general pathology," 207.

36. Sigerist, *Man and Medicine*, 215.

37. Foster, *A History of Medical Bacteriology and Immunology;* Gieson, *The Private Science of Louis Pasteur;* Worboys, *Spreading Germs;* and Gradmann, *Laboratory Disease.*

38. Alfred B. Garrod, *A Treatise on Gout and Rheumatic Gout (Rheumatoid Arthritis),* and "The great practical importance of separating rheumatoid arthritis from gout"; and Charcot, *Clinical Lectures on Senile and Chronic Diseases.* See also Bywaters, "Historical aspects of the aetiology of rheumatoid arthritis"; and Storey, Comer, and, Scott, "Chronic arthritis before 1876." Confusion with osteoarthritis was common until the early twentieth century: for their differentiation, see Nicholls and Richardson, "Arthritis deformans."

39. Alfred B. Garrod, "Great practical importance," 1036.

40. Archibald E. Garrod, *A Treatise on Rheumatism and Rheumatoid Arthritis,* and "A contribution to the theory of the nervous origin of rheumatoid arthritis."

41. Poynton and Paine, "The etiology of rheumatic fever." See also Benedek, "The history of bacteriologic concepts of rheumatic fever and rheumatoid arthritis."

42. Hunter, *Oral Sepsis*; Billings, "Chronic focal infections"; and Billings, Coleman, and Hibbs, "Chronic infectious arthritis."

43. Louie, "Renoir, his art, and his arthritis"; and Boonen, Rest, Dequeker, and Linden, "How Renoir coped with rheumatoid arthritis."

44. Cazenave, "Lupus érythèmateux (érythème centrifuge)." See also Douglas and Cyr, "The history of lupus erythematosus from Hippocrates to Osler"; Potter, "The history of the disease called lupus"; Wallace and Lyon, "Pierre Cazenave and the first detailed modern description of lupus erythematosus."

45. Kaposi, "Neue Beiträge zur Kenntniss des Lupus erythematosus."

46. Osler, "On the visceral complications of erythema exudativum multiforme," and "On the visceral manifestations of the erythema group of skin diseases." See also Benedek, "Historical background of discoid and systemic lupus erythematosus"; and Mallavarapu and Grimsley, "The history of lupus erythematosus."

47. Hutchinson, "Harveian lectures on lupus."

48. Lancereaux, "Le diabète maigre." See also Gale, "The discovery of type I diabetes," and "The rise of childhood type I diabetes."

49. Himsworth, "Diabetes mellitus." See also Feudtner, *Bittersweet.*

50. Morse, "Diabetes in infancy and childhood"; and Gundersen, "Is diabetes of infectious origin?"

51. Edwards, "Clinical memoranda," 279.

52. Wilson, "Diabetes in a young child."

53. Anderson, "Progress of medical science," 248.

54. Rachford, "A case of diabetes mellitus."

55. Firth, "The case of Augustus d'Este." See also Frederickson and Kam-Hansen, "The 150-year anniversary of multiple sclerosis"; and Quétel, *History of Syphilis*.

56. Jellinek, "Heine's illness"; and Stenager, "The course of Heinrich Heine's illness."

57. Carswell, *Pathological Anatomy*; and Cruveilhier, *Anatomie pathologique du corps humaine*. Carswell had studied with Pierre Charles and Alexandre Louis in Paris.

58. Charcot, "Histologie de la sclèrose en plaques." See also Hickey, "The pathology of multiple sclerosis"; Talley, "The emergence of multiple sclerosis," and *A History of Multiple Sclerosis*; and Murray, *Multiple Sclerosis,* and "The history of multiple sclerosis." In the United States, Surgeon General William Alexander Hammond, informed by his knowledge of Charcot's work, soon discussed nine cases of multiple sclerosis in his *Treatise on Diseases of the Nervous System*. See Blustein, *Preserve Your Love for Science*. The first confirmed English case is described in Moxon, "Case of insular sclerosis." Talley attributes the apparently increased prevalence of multiple sclerosis mostly to better neurological training and physician vigilance: see his "The emergence of multiple sclerosis."

59. Marie, "Sclèrose en plaques." Hammond also regarded the disease as infectious.

60. Barbellion, *The Journal of a Disappointed Man*, 5 July 1917, 292.

61. Wells, "Introduction," vi, viii.

62. Barbellion, *Journal*, 26 April 1913, 80; and 16 February 1914, 109.

63. Barbellion, *Journal*, 4 March 1915, 180; and 24 March 1915, 182.

64. Barbellion, *Journal*, 8 November 1915, 224; 28 November 1916, 258; and 1 November 1916, 255.

65. Barbellion, *Journal*, 5 March 1917, 285 and 305. The book stated that Barbellion died on 31 December 1917, but Cummings did not die until 22 October 1919, so he lived to see his journal published in March 1919.

Chapter 2. Immunological Thought Styles

1. Lister, "The relations of clinical medicine," 733, 734. See also Lawrence and Dixey, "Practising on principle."

2. Gradmann, *Laboratory Disease*.

3. Dubos, *Louis Pasteur*; and Geison, *The Private Science of Louis Pasteur*. On Pasteur's biological orientation, see Mendelsohn, "'Like all that lives'"; and Weindling, "Scientific elites and laboratory organization."

4. Bulloch, *History of Bacteriology*; Foster, *A History of Medical Bacteriology and Immunology*; Silverstein, *A History of Immunology*; and Moulin, *Le dernier langage de la médecine*.

5. Virchow, "Recent progress in science and its influence on medicine and surgery," 243.

6. Wells, *The War of the Worlds*, 187.

7. Cohen, *A Body Worth Defending.*

8. Metchnikoff, "A yeast of Daphnia," 195. See also Metchnikoff, "Concerning the relationship between phagocytes and anthrax bacilli," and *Lectures on the Comparative Pathology of Inflammation.* On Metchnikoff, see Chernyak and Tauber, "The idea of immunity"; and Tauber and Chernyak, *From Metaphor to Theory.*

9. Smith, 1915, quoted in Arthur F. Coca, Historical Summary of the Origin of the *Journal of Immunology* [typescript 1950], in Edsall, "What is immunology?" 167. An allergist, Coca was the founding editor of the journal. Smith had demonstrated some decades earlier that a dead virus could confer protection against the living form: Salmon and Smith, "On a new method of inducing immunity."

10. Ehrlich, "On immunity with special reference to cell life," 178, 179. See also Marquardt, *Paul Ehrlich*; and Silverstein, *Paul Ehrlich's Receptor Immunology.*

11. Ehrlich, "On immunity," 180.

12. Ehrlich, "On immunity." See also Silverstein, *History of Immunology.* Jules Bordet had already discovered this additional protective substance causing bacteriolysis and hemolysis (when red blood cells are part of the immune complex) in 1898, calling it alexine. Max von Gruber and Herbert Dunham had observed immune serum causing clumping or agglutinating of bacteria in 1896.

13. Crist and Tauber, "Debating humoral immunity and epistemology." On debates over affinity and avidity, used as proxies for degree of specificity, see Keating, Cambrosio, and Mackenzie, "The tools of the discipline."

14. Arrhenius, *Immunochemistry.*

15. Mazumdar, *Species and Specificity;* and Silverstein, *History of Immunology.* Mazumdar stresses the influence on Gruber of the evolutionary biologist Carl von Nägeli, a believer in bacterial variability. On Pettenkoffer and contingent contagionism, see Winslow, *The Conquest of Epidemic Disease;* Rosenberg, *The Cholera Years;* and Pelling, *Cholera, Fever and English Medicine.*

16. Mazumdar, "The purpose of immunity." Landsteiner received the 1930 Nobel Prize in Physiology or Medicine for his discovery of human blood groups.

17. Mackenzie, "Paroxysmal haemoglobinuria, with remarks on its nature."

18. Donath and Landsteiner, "Über paroxysmale Hämoglobinurie," and "Weitere Beobachtungen über paroxysmale Hämoglobinurie." See Silverstein, *History of Immunology*, Chapter 8.

19. Ehrlich and Morgenroth, "Fifth communication on hemolysis," 255. Ehrlich was responding to Serge Metalnikoff's discovery in Metchnikoff's laboratory of antibodies against spermatozoa ("Études sur la spermotoxine"). See Silverstein, "Horror Autotoxicus, Autoimmunity, and Immunoregulation."

20. Wassermann, Neisser, Bruck, and Schucht, "Weitere Mitteilungen über den Nachweis spezifischluetischer Substanzen durch Komplementverankerung." This is a complement fixation test used to detect antibodies. See Bialynicki-Birula, "The 100th anniversary of the Wasserman-Neisser-Bruck reaction."

21. Landsteiner, Müller, and Poetzl, "Zur Frage der Komplementbindungsreaktionen bei Syphilis." Typically, Landsteiner later speculated that the antibody might be responding to chemically similar substances in spirochetes and animal tissues; see his *Die Spezifizität der Serologischen Reaktionen*. Landsteiner also showed interest in pemphigus, later explained as an autoimmune disease; see Landsteiner, Levaditi, and Prásek, "Étude expérimentale du pemphigus infectieux aigu."

22. Weil and Braun, "Ueber das Wesen der luetischen Erkrankung auf Grund der neueren Untersuchungen"; and Weil, "Das Problem der Serologie der Lues in der Darstellung Wassermanns." In 1915, Weil and Arthur Felix developed the Weil-Felix test for typhus and rickettsial diseases. Weil also influenced Anton Elschnig's studies in Prague of sympathetic ophthalmia, which concluded that hypersensitivity in the lens after trauma caused greater immunological reactivity against the other eye (Elschnig, "Studien zur sympatischen Ophthalmie").

23. Lesky, *The Vienna Medical School of the Nineteenth Century*. In 1895, Virchow criticized the double aspect of Viennese pathology: "On the one hand, the strict, natural-scientific pathological-anatomical depiction of disease; on the other, speculative constructions in basic pathology forming all sorts of arbitrary patterns" ("One hundred years of general pathology," 197). In particular, he disliked Carl Rokitansky's "false pathology of the humors" (205), with its emphasis on the chemical composition of the blood in disease.

24. While Kraus has been associated with Ehrlich, he did not train in Berlin and remained skeptical of arguments for absolute specificity of immunological reactions. Kraus discovered precipitin, a substance causing precipitation of antigen-antibody complexes in the blood. He spent 1912–23 in Argentina and Brazil, before returning to Vienna. Kraus and Wassermann together founded the Free Association for Microbiology. See Lesky, "Wassermann and the Vienna school of serology," and "Viennese serological research about the year 1900"; and Schick, "Pediatrics in Vienna at the turn of the century."

25. Schorske, *Fin-de-Siècle Vienna*, 10. See also Janik and Toulmin, *Wittgenstein's Vienna*; Beller, *Vienna and the Jews, 1867–1938*; Coen, "Living precisely in fin-de-siècle Vienna"; Buklijas, "Surgery and national identity in late nineteenth-century Vienna"; and Hacohen, "The culture of Viennese science and the riddle of Austrian liberalism."

26. Metchnikoff, *The Prolongation of Life*. His claims were prompted by Bouchard, *Lectures on Autointoxication*.

27. Richet and Portier, "De l'action anaphylactique de certains venins"; and Richet, *L'Anaphylaxie*. Bordet regarded anaphylaxis as an "accident in the course

of the defense" ("Anaphylaxis—its importance and mechanism," 464). In 1903, Nicolas-Maurice Arthus discovered the basis of local hypersensitivity reactions ("Injections répétées de serum du cheval chez le lapin").

28. Carroy points out that Richet was "often at the forefront of modernity in various forms" ("Playing with signatures: the young Charles Richet," 245). See Löwy, "On guinea pigs, dogs and men"; Kroker, "Immunity and its other"; and Jackson, *Allergy.*

29. Richet, "Ancient humoralism and modern humoralism," 921, 924, 925. In his 1913 Nobel Prize lecture, Richet coined the term *humoral personality* ("Anaphylaxis").

30. Pirquet and Schick, *Serum Sickness.* See also Schick, "Pediatrics in Vienna."

31. Pirquet and Schick, *Serum Sickness,* 119.

32. Pirquet, *Allergy,* 55. This expands his earlier publication: "Allergie." William H. Welch briefly recruited Pirquet to the Johns Hopkins Medical School but he soon returned to Vienna. In 1929 he and his wife took cyanide and died.

33. The English pharmacologist Henry Dale resolved disputes about the mechanism of allergy when he recognized that it was mediated by the release of histamine, which he had discovered in 1910; see Dale, "The biological significance of anaphylaxis." See also Jackson, *Allergy*; and Mitman, *Breathing Space.*

34. Shaw, "Preface on the doctors," 28.

35. Temkin, "The scientific approach to disease."

36. A. E. Garrod, "The incidence of alkaptonuria." Later, Garrod asserted, "What used to be spoken of as a diathesis is nothing else but chemical individuality" (*Inborn Factors in Disease,* 57). See Burgio, "Biological individuality and disease"; Moulin, "The dilemma of medical causality and the issue of biological individuality"; and Mendelsohn, "Medicine and the making of bodily inequality in twentieth-century Europe."

37. Starling, "The wisdom of the body," 689. Earlier Starling had identified secretin (Bayliss and Starling, "The mechanism of pancreatic secretion"). See Medvei, *The History of Clinical Endocrinology.*

38. Funk, "The effect of a diet of polished rice on the nitrogen and phosphorus of the brain." See also Apple, *Vitamania: Vitamins in American Culture.*

39. Adler, "On the neurotic disposition," 98, 103. See Parnes, " 'Trouble from within.' "

40. Hirszfeld, "Ueber die Konstitutionsserologie im Zusammenhang mit der Blutgruppenforschung." Hirszfeld studied at Würzburg and Berlin before World War I; after the war he became director of the State Hygiene Institute in Warsaw. He was incarcerated in the Warsaw ghetto during World War II. See Keating, "Holistic bacteriology"; and Löwy, "Immunology in the clinics." For similar claims later made about the tissue antigens, see Chapter 6.

41. Leo Loeb, *The Biological Basis of Individuality*, vii. Loeb wrote the first draft in 1930. See also Leo Loeb, "The biological basis of individuality." Loeb's interests included cell culture and tissue transplantation.

42. For example, see Vallery-Radot and Heimann, *Hypersensibilités specifiques dans les affections cutanées*, which reflects the influence in France of Claude Bernard, Charles Richet, and Fernand Widal. Also in the 1930s, Hans Selye developed theories about the adaptive response to stress, mediated through the immune system; see Selye, "Allergy and the general adaptation syndrome"; and Viner, "Putting stress in life."

43. Jackson, "John Freeman, hay fever and the origins of clinical allergy in Britain."

44. In Lewis's novel, Max Gottlieb encourages Martin Arrowsmith to treat all living creatures as "physicochemical machines" (Lewis, *Arrowsmith*, 28). See also Rosenberg, "Martin Arrowsmith"; and Löwy, "Immunology and literature in the early twentieth century." Jacques Loeb (brother of Leo) wrote *The Mechanistic Conception of Life*. See Pauly, *Controlling Life*.

45. Arthur F. Coca to Hans Zinsser, 21 February 1928, folder 9, box 2, Hans Zinsser papers, HMS c73, Countway Medical Archives, Harvard University. Coca, Zinsser, and Landsteiner were good friends; the three met regularly.

46. Corner, *A History of the Rockefeller Institute*; and Stapleton, ed., *Creating a Tradition of Biomedical Research*.

47. Swift, "Rheumatic fever," 2078. In his 1933 annual report, Swift noted, "The disease is due to streptococci acting in tissues that are altered in their mode of reactivity either as a result of infection or other factors" (Rockefeller Institute Scientific Reports, 21 [1931–33], 208–26, quote on 208, Record Group [RG] 439, box 4, Rockefeller University, Rockefeller Archive Center [RAC]). See also Swift, "The nature of rheumatic fever." Rebecca Lancefield in Swift's laboratory devised the serological classification of streptococci. On Swift's retirement, Maclyn McCarty took over his work on streptococcal infection.

48. Zinsser, *As I Remember Him*, 47. See also Summers, "Hans Zinsser." Robert Koch used tuberculin for treatment of tuberculosis in 1890 but it proved ineffective. Later it was used to diagnose previous exposure to tuberculosis germs.

49. Zinsser, Ward, and Jennings, "The significance of bacterial allergy as a sign of resistance," 723.

50. Zinsser, "On the significance of bacterial allergy in infectious diseases," 381.

51. Benison, *Tom Rivers*, 92. During his early years at the Rockefeller Institute, Rivers studied chicken pox, psittacosis, and louping-ill, a disease of sheep analogous to polio. "I never did make up my mind about the question of immunity to psittacosis," he later reflected (Benison, *Tom Rivers*, 161).

52. Benison, *Tom Rivers*, 174. See Rivers, Sprunt, and Berry, "Observations on attempts to produce acute disseminated encephalomyelitis in monkeys"; Schwentker

and Rivers, "The antibody response of rabbits to injections of emulsions of homologous brain"; and Rivers and Schwentker, "Encephalomyelitis accompanied by myelin destruction experimentally produced in monkeys." Another Hopkins medicine graduate, Schwentker had worked as a pediatrician in Baltimore. During World War II, he assisted Rivers at the Rockefeller Institute Naval Institute and the Naval Medical Research Institute No. 2 on Guam. He was professor of pediatrics at Hopkins 1945–54.

53. Rivers, "Viruses," 1152.

54. Benison, *Tom Rivers*, 174. Rivers went on to observe, "The work I did on demyelinating encephalitis is probably the nicest piece of work I ever did" (175). See Mackay and Anderson, "What's in a name?"

55. Smadel, "Experimental nephritis in rats induced by injections of anti-kidney serum preparation"; and Smadel and Swift, "Reverse anaphylaxis in rats." Japanese pathologist Matazo Masugi had earlier induced nephritis in rats by injecting a rabbit antiserum prepared against rat kidney tissue ("Zur Pathogenese der Diffusen Glomeronephritis als allergischer Erkrankung der Niere").

56. H. F. Swift, Rockefeller Institute Scientific Reports, 24 (1935–36), 223–33, 227, RG 439, box 4, Rockefeller University, RAC.

57. H. F. Swift, Rockefeller Institute Scientific Reports, 25 (1936–37), 214–24, 221, RG 439, box 4, Rockefeller University, RAC.

58. Schwentker and Comploir, "The production of kidney antibodies by injection of homologous kidney," 229, 230.

59. In 1937, Rivers replaced Rufus Cole as director of the Rockefeller Hospital. After World War II, Smadel worked at Walter Reed Hospital and became associate director for intramural research at the National Institutes of Health, where he became a mentor of D. Carleton Gajdusek (see Chapter 4).

60. Landsteiner, *The Specificity of Serological Reactions.*

61. Fleck, "Some specific features of the medical way of thinking." Fleck also was committed to Hirszfeld's theories of constitutional serology or biochemical idiosyncrasy; see Fleck, "Sérologie constitutionelle." Löwy attributes Fleck's ideas to his connection to Hirszfeld and his involvement with other Polish philosophers of medicine; see *The Polish School of Philosophy of Medicine.*

62. Fleck, *Genesis and Development of a Scientific Fact*, 52–53. See also Moulin, "Fleck's style"; Löwy, "The epistemology of the science of an epistemologist of the sciences," and "The immunological constitution of the self"; and Belt and Gremmen, "Specificity in the era of Koch and Ehrlich." Fleck conducted research on lupus and pemphigus, two diseases later discovered to be autoimmune. See Fleck and Füllenbaum, "Clinical and experimental contribution on etiology of lupus erythematosus"; Fleck and Goldschlag, "Experimentelle Beiträge zur Pemphigusfrage"; and "Further experimental studies of pemphigus." In 1956 he attended a congress on autoimmunity at the University of Texas.

63. Fleck, *Genesis and Development*, 65, 110.

Chapter 3. A Sense of Unlimited Possibilities

1. Sullivan, "The new scientific horizon," 722. See Bradshaw, "The best of companions: J. W. N. Sullivan, Aldous Huxley, and the new physics." Sullivan wrote biographies of Newton, Einstein, and Beethoven. No doubt Aldous Huxley also learned much biological science from his brother Julian.

2. Virginia Woolf, 8 December 1921, in *The Diary of Virginia Woolf*, ed. Bell and McNellie, vol. 2, 150.

3. Turner, *The Duchess of Popocatapetl*, 248.

4. Marks, *The Progress of Experiment.*

5. In 1930, Kathleen Chevassut claimed to have isolated the germ of multiple sclerosis, but the claim was soon discredited ("The aetiology of disseminated sclerosis"). See Caspar, "Trust, protocol, gender, and power."

6. Finlay, "Pathogenesis of encephalitis occurring with vaccination, variola and measles"; Kennedy, "Allergy and its effects on the central nervous system"; and Winkelman and Moore, "Allergy and nervous diseases."

7. Ferraro, "Pathology of demyelinating diseases as an allergic reaction of the brain," 448. See also Ferraro and Kilman, "Experimental toxic approach to mental disease." Ferraro is most often remembered for identifying brain damage from electro-convulsive therapy, which had recently been introduced at the Psychiatric Institute. He was also interested in the effects of vitamin and nutritional deficiencies on the brain. In his retirement, he became a vehement critic of psychoanalysis.

8. Ferraro and Jervis, "Experimental disseminated encephalopathy in the monkey"; and Ferraro, "Allergic brain changes in post-scarlatina encephalitis."

9. Ferraro, "Pathology of demyelinating diseases." See also Stevenson and Alvord, "Allergy in the nervous system."

10. Politically progressive and a dabbler in psychoanalysis, Putnam had studied medicine at Harvard before completing his internship at Johns Hopkins and later training with Harvey Cushing. His uncle, James Jackson Putnam, was the Harvard neurologist who invited Sigmund Freud to lecture in the United States. Tracy Putnam led the excellent neurological team at Boston City Hospital, where he helped discover the benefits of phenytoin (Dilantin) in epilepsy, before coming to New York in 1939. He left the Neurological Institute in 1947, moving to private practice in Los Angeles, where he experienced financial difficulties and played the role of a physician in the horror film *The Slime People*. See Rowland, *The Legacy of Tracy J. Putnam and H. Houston Merritt.*

11. Putnam, McKenna, and Morrison, "Studies in multiple sclerosis. I: The histogenesis of experimental sclerotic plaques and their relation to multiple sclerosis"; and Putnam, "The pathogenesis of multiple sclerosis," "Studies in multiple sclerosis. IV: 'Encephalitis' and sclerotic plaques produced by venular obstruction," "Studies in multiple sclerosis. VIII: Etiological factors in multiple sclerosis," and "Venous thrombosis as the primary alteration in the lesions of 'encephalomyelitis'

and multiple sclerosis." See also Talley, *A History of Multiple Sclerosis;* and Murray, *Multiple Sclerosis.*

12. Ferraro, "Pathology of demyelinating diseases." See also Hurst, "A review of some recent observations on demyelination."

13. Putnam continued to treat multiple sclerosis patients with anticoagulants, or blood "thinning" agents, until the 1970s. See Talley, "The treatment of multiple sclerosis in Los Angeles and the United States, 1947–1960."

14. In 1913, the Highlanders became the Yankees, and ten years later the team moved to the Bronx.

15. Harvey, *Science at the Bedside;* and Maulitz and Long, eds., *Grand Rounds.*

16. Kabat, "Getting started 50 years ago," 13. Heidelberger had worked with Landsteiner at the Rockefeller Institute in the 1920s.

17. Kabat, Wolf, and Bezer, "The rapid production of acute disseminated encephalomyelitis." Kabat and colleagues had issued a brief preliminary note on their research: Kabat, Wolf, et al., "Rapid production of acute disseminated encephalomyelitis in rhesus monkeys."

18. Morgan, "Allergic encephalomyelitis in monkeys in response to injection of normal monkey nervous tissue." The daughter of biologist Thomas Hunt Morgan and a graduate of the University of Pennsylvania, Isabel Morgan (later Morgan Mountain) worked at the Rockefeller Institute from 1938 until 1944, when she left for Hopkins to conduct research on killed-virus polio vaccines. In 1949 she married and went to New York, abandoning polio research.

19. Kabat, Wolf, and Bezer, "The rapid production of acute disseminated encephalomyelitis," 127. In 1949, Kabat used autologous brain tissue from monkeys as the immunogen, proving that the encephalitis was truly auto-allergic: Kabat, Wolf, and Bezer, "Studies on acute disseminated encephalomyelitis produced experimentally in rhesus monkeys. IV: Disseminated encephalomyelitis produced in monkeys with their own brain tissue." Both Putnam and Ferraro had earlier commented on the similarities of experimental encephalomyelitis and multiple sclerosis.

20. Dienes, "The immunological significance of the tuberculous tissue." Freund worked with fellow Hungarian Dienes at the Von Ruck Research Laboratory for Tuberculosis in Asheville, North Carolina (1924–26); Dienes had been a protégé of the Prague bacteriologist Edmund Weil during World War I.

21. Freund and McDermott, "Sensitization to horse serum by means of adjuvants." See also Freund, "The effect of paraffin oil and mycobacteria on antibody formation and sensitization"; Freund, Stern, and Pisani, "Isoallergic encephalomyelitis and radiculitis in guinea pigs"; and Freund, Lipton and Thompson, "Aspermatogenesis in the guinea pig induced by testicular tissue and adjuvants." In his analysis of the discovery of the adjuvant, Nicolas Rasmussen separates technical innovation from conceptual change in immunology, but we are implying that Freund's interest in immunological sensitization derived from his underlying commitment to a particular "allergic" thought style. See Rasmussen, "Freund's adjuvant and the

realization of questions in postwar immunology." Freund later became the chief of the immunology laboratory at the Institute of Allergy and Infectious Diseases in Bethesda, Maryland.

22. Burnet, *Cellular Immunology*, 593.

23. For example, Kabat, Wolf, and Bezer, "Studies on acute disseminated encephalomyelitis produced experimentally in rhesus monkeys. VII: The effect of cortisone."

24. Kabat, "Getting started 50 years ago," 32. Kabat remained at Columbia University the rest of his career. Baruj Benacerraf recalled that Kabat was "mercilessly opposed to loose thinking and idle speculation. Devastating criticism, voiced in a characteristically high-pitched, rasping, loud tone, heard throughout the whole floor, were directed to the unfortunate one breaking any of these rules" ("Reminiscences," 8). Noel Rose remembered Kabat as "very strict and a very good guide—but he was not an easy person to get along with." Claiming to be allergic to tobacco smoke, "he used to come to immunology meetings wearing a gas mask—he was not shy about expressing himself" (Noel Rose, interviewed by Warwick Anderson, 30 April 2012, Baltimore, MD).

25. Morrison, "Disseminated encephalomyelitis experimentally produced by the use of homologous antigen"; and Freund, Lipton, and Morrison, "Demyelination in the guinea pig in chronic allergic encephalomyelitis." Morrison died in 1950, aged 53.

26. After completing his M.D. at the University of Pennsylvania, Waksman became fascinated by rheumatic fever while training at the Mayo Clinic, leading him to study immunology in New York. See Waksman, "The etiology of rheumatic fever."

27. Waksman and Morrison, "Tuberculin type sensitivity to spinal cord antigen in rabbits with allergic encephalomyelitis"; and Waksman, "Further studies of skin reactions in rabbits with experimental allergic encephalomyelitis," and "Experimental allergic encephalomyelitis and the 'auto-allergic' diseases." Waksman later worked at Yale and then became director of research in the National Multiple Sclerosis Society.

28. Evans and Harding, "Multiple sclerosis," 280. See also Robinson, "Personal narratives, social careers and medical courses"; and Monks and Frankenburg, "Being ill and being me."

29. Flannery O'Connor to Robert and Elizabeth Lowell, 17 March 1953, in O'Connor, *The Habit of Being*, 57.

30. Flannery O'Connor to Elizabeth Fenwick Way, 4 August 1957, in O'Connor, *The Habit of Being*, 233.

31. Hargraves, Richmond, and Morton, "Presentation of two bone marrow elements: The 'tart' cell and the 'L.E.' cell"; and Hargraves, "Discovery of the L.E. cell and its morphology." On "team research" at the Mayo Clinic, see Harvey, "Clinical science at the Mayo Clinic."

32. Haserick, Lewis and Bortz, "Blood factor in acute disseminated lupus erythematosus."

33. Miescher and Fauconnet, "L'absorption du facteur 'L.E.' par des noyaux cellulaires isolés."

34. Coons and Kaplan, "Localization of antigen in tissue cells." Coons studied with Hans Zinsser and John Enders.

35. Holman and Kunkel, "Affinity between the lupus erythematosus serum factor and cell nuclei and nucleoprotein," 163. We are grateful to Hal Holman for clarifying the pattern of experiments (Holman email to Warwick Anderson, 9 January 2014). The son of Stanford medical school professors, Holman spent much of the 1950s campaigning against nuclear weapons; and in the 1960s he opposed the Vietnam War, even while chairing the Department of Medicine at Stanford. Around 1957, George J. Friou at Yale used immunofluorescence to reveal the same phenomenon ("Clinical application of lupus serum nucleoprotein reaction using fluorescent antibody technique"). W. C. Robbins led experiments that found complement-fixing antibodies to nuclear components of the cell, especially DNA, in SLE sera: Robbins, Holman, Deicher, and Kunkel, "Complement fixation with cell nuclei and DNA in lupus erythematosus."

36. Tan and Kunkel, "Characteristics of a soluble nuclear antigen"; and Friou, "Antinuclear antibodies."

37. Halstead R. Holman interviewed by Warwick Anderson, 12 November 2012, Stanford, CA.

38. Halstead R. Holman interviewed by Warwick Anderson, 12 November 2012, Stanford, CA.

39. Theodor Svedberg developed the ultracentrifuge in the 1920s but commercial versions were not available until the 1940s; see Elzen, "Two ultracentrifuges." Arne Tiselius invented electrophoresis in 1937; see Kay, "Laboratory technology and biological knowledge."

40. Halstead R. Holman interviewed by Warwick Anderson, 12 November 2012, Stanford, CA.

41. Aladjem and Schur, *In Search of the Sun*, 29, 100. Portions of this book derive from Aladjem, *The Sun Is My Enemy.*

42. Aladjem and Schur, *In Search of the Sun*, 152–53. Aladjem was a founder of the Lupus Foundation of America and edited *Lupus News.*

43. Klemperer, Pollack, and Baehr, "Diffuse collagen disease"; and Klemperer, "The pathogenesis of lupus erythematosus."

44. Klemperer, "The concept of collagen diseases," 505, 515.

45. Waaler, "On the occurrence of a factor in human serum." See Fraser, "The Waaler-Rose test."

46. Rose, Ragan, Pearce, and Lipman, "Differential agglutination of normal and sensitized sheep erythrocytes." It is certain that at least Lipman had attended Waaler's 1939 talk.

47. Franklin, Holman, Müller-Eberhard, and Kunkel, "An unusual protein component of high molecular weight in the serum of certain patients with rheumatoid arthritis"; and Kunkel and Ward, "Clinical and laboratory findings in seven patients with rheumatoid arthritis."

48. Kunkel, "Accomplishments," vi. These notes were written the day before Kunkel died. We are grateful to Shu-Man Fu for drawing our attention to them.

49. Fudenberg and Franklin, "Rheumatoid factors and the etiology of rheumatoid arthritis."

50. See Kay, *The Molecular Vision of Life;* and Chadarevian and Kamminga, eds., *Molecularizing Biology and Medicine.* Kunkel encouraged Gerald Edelman, his Ph.D. student, to investigate the chemical structure of globulins, for which Edelman shared the 1972 Nobel Prize in Physiology or Medicine.

51. For example, Avery, MacLeod, and McCarty, "Induction of transformation by a desoxyribonucleic acid fraction"; Hershey and Chase, "Independent functions of viral protein and nucleic acid in growth of bacteriophage"; Watson and Crick, "Molecular structure of nucleic acids"; Crick and Watson, "Structure of small viruses"; and Lwoff, "The concept of the virus."

52. Vesell, "Recollections from the Kunkel laboratory," 239. See also Robbins, "Recollections of Henry G. Kunkel." Kunkel's father was a senior plant pathologist at the Rockefeller Institute, so he understood the institute's culture well.

53. Benacerraf, "Henry G. Kunkel, 1916–1983," 878.

54. Said Holman: "We were dealing with certain elements of cellular function and . . . the things we studied were mostly antibodies. Now Gerry Edelman came into the lab and he worked on the structure of rheumatoid factor, so Gerry went in the direction of basic chemistry. But others of us didn't go that way" (Halstead R. Holman interviewed by Warwick Anderson, 12 November 2012, Stanford, CA).

55. We discuss autoimmune hepatitis in Chapter 4.

56. Mackay and Burnet, *Autoimmune Diseases*, ix. See Mackay, "The 'Autoimmune Diseases' 40th anniversary." The major omissions, as Mackay points out, were type 1 diabetes mellitus (see Chapter 5), pemphigus, alopecia, vitiligo, psoriasis, and celiac disease.

57. Dameshek and Schwartz, "The presence of hemolysins in acute hemolytic anemia"; and Dameshek and Schwartz, with Gross, "Hemolysins as the cause of clinical and experimental hemolytic anemias."

58. Coombs, Mourant, and Race, "A new test for detection of weak or 'incomplete' Rh agglutinins." Coombs later acknowledged that Carlo Moreschi had earlier devised a similar test ("Neue tatsachen über die Blutkörperchenagglutination"). A veterinary graduate, Coombs later concentrated on the classification of allergic reactions.

59. Young, Miller, and Christian, "Clinical and laboratory observations on autoimmune hemolytic disease." See also Fudenberg and Kunkel, "Physical properties of the red cell agglutinins in acquired hemolytic anemia." J. V. Dacie wrote

in 1962, "During the last 10 years or so the antibodies of acquired haemolytic anaemia have been extensively studied and the concept of auto-immunization as a cause of antibody formation has been widely accepted" (*The Haemolytic Anaemias*, 345). For the discovery of the autoimmune basis of another blood disease, the clotting disorder idiopathic thrombocytopenic purpura, see Harrington, Minnich, Hollingsworth and Moore, "Demonstration of a thrombocytopenic factor in the blood of patients with thrombocytopenic purpura."

60. Noel R. Rose, "The discovery of thyroid autoimmunity," 167. See also Noel R. Rose, "Autoimmunity," and "Thyroid autoimmunity." Rose had completed his Ph.D. at Penn and was studying for his M.D. at Buffalo.

61. Rose recounted that Freund had showed him how to make the adjuvant in a whisky glass: "You keep emulsifying until you get a drop that floats in a beaker of water. I still teach my students the same thing. If you don't do that—and many people didn't—you get antibodies but you don't get lesions" (Noel Rose, interviewed by Warwick Anderson, 30 April 2012, Baltimore, MD).

62. Rose, "The discovery of thyroid autoimmunity," 168. See Witebsky and Rose, "Studies on organ specificity. IV: Production of rabbit thyroid antibodies in the rabbit"; Rose and Witebsky, "Studies on organ specificity. V: Changes in thyroid glands of rabbits following active immunization with rabbit thyroid extracts"; and Beutner, Witebsky, Rose, and Gerbasi, "Localization of thyroid and spinal cord autoantibodies by fluorescent antibody technic."

63. Witebsky, Rose, Terplan, Paine, and Egan, "Chronic thyroiditis and autoimmunization." Rose and colleagues used tanned cell hemagglutination, in which red blood cells are treated with tannic acid to absorb soluble antigens, so they will agglutinate in the presence of the specific antibody.

64. Ivan Roitt, interviewed by Warwick Anderson, 3 June 2013, Golders Green, London.

65. Roitt, Doniach, Campbell, and Hudson, "Auto-antibodies in Hashimoto's disease (lympadenoid goitre)"; and Doniach and Roitt, "Autoimmunity in Hashimoto's disease and its implications." See also Doniach and Roitt, "Human organ specific immunity." Doniach was working as a medical assistant in a thyroid clinic; Roitt had studied chemistry at Oxford. Among the organ-specific autoimmune diseases later identified are pernicious anemia, Addison's disease, chronic active hepatitis, and type 1 diabetes mellitus.

66. Noel Rose, interviewed by Warwick Anderson, 30 April 2012, Baltimore, MD.

67. Ivan Roitt, interviewed by Warwick Anderson, 3 June 2013, Golders Green, London. Doniach and Roitt went on to determine that pernicious anemia and primary biliary cirrhosis are autoimmune diseases. When Doniach and Roitt visited Rose, they found Witebsky "a bit dismissive of us. He thought we were upstarts and knew very little." In contrast, in New York they played piano four

hands accompanied by Heidelberger on clarinet, and even enjoyed conversation with Kabat.

68. Anon., "The immunology of thyroid disease"; and Anon., "Autoimmunity in thyroid disease."

69. Witebsky, "Historical roots of present concepts of immunopathology," 9. Witebsky intensely disliked Landsteiner for having snubbed him in New York, and Heidelberger and Kabat for their chemical orientation (Noel Rose, interviewed by Warwick Anderson, 30 April 2012, Baltimore, MD). According to Rose, Witebsky became "concerned that our findings, casting doubt on the doctrine of *horror autotoxicus*, might lead to a flood of articles incriminating autoimmunity in many human diseases of unknown etiology" ("Life among the contrivances," 1011). See also Rose and Bona, "Defining criteria for autoimmune diseases." Rose became director of the Research Center for Autoimmunity at the Johns Hopkins University.

70. Witebsky, et al., "Chronic thyroiditis and autoimmunization," 1446.

71. Witebsky, "Historical roots," 10. Despite his misgivings, Witebsky continued to make major contributions to research in autoimmunity.

72. Noel Rose, interviewed by Warwick Anderson, 30 April 2012, Baltimore, MD.

73. A few also stand out for their radical political commitments, especially Kabat and Holman, who in the Bay Area became an advocate of action research and lost his job at Stanford during the Vietnam War. See Holman, "The crisis technology poses for political institutions."

74. On the contribution of such new "epistemic things" to experimental systems, see Rheinberger, *Toward a History of Epistemic Things.*

75. Jackson, *Allergy,* Chapter 3. There was also a discomforting appropriation of allergy in discussions of personality types and constitutional predispositions: see for example, Selye, "Allergy and the general adaptation syndrome."

76. Mackay and Anderson, "What's in a name?" Some immunologists continued to favor *allergy*, believing it made more sense biologically; for example, Robin Coombs at a Wellcome Trust Witness Seminar in 1995 declared: "I can't bear this use of autoimmunity, and I shall go down fighting it to the end" (in *Self and Non-Self*, ed. Tansey, Willhof, and Christie, 47).

Chapter 4. The Science of Self

1. Burnet, "Medical education and research," 557, 558, 561.

2. Notes on American Visit, 1943–44, folder 11a, box 8, series 2, F. Macfarlane Burnet papers 1989/0034, University of Melbourne Archives [Burnet papers].

3. Burnet, *Walter and Eliza Hall Institute*, 56.

4. Burnet, *Changing Patterns*, 73. Burnet writes: "I wanted to bring medical research in Australia into a form patterned on the American model" (73).

5. Notes on American visit, 1943–44, folder 11, box 8, series 2, Burnet papers.

6. October 1950, Burnet Travel Diary 1950, folder 14, box 8, series 2, Burnet papers.

7. October 27, 1952, American Travel Diary 1952, folder 17, box 8, series 2, Burnet papers.

8. October 18, 1954, Travel and General, 1954, folder 19, box 8, series 2, Burnet papers.

9. For multiple examples, see his General Diary, 29 January 1922–17 April 1924, folder 7, series 2, Burnet papers. Burnet was then in his early twenties, working as a pathologist at the Melbourne Hospital.

10. Burnet, "The basis of allergic diseases," 33, 29.

11. Burnet, "Basis of allergic diseases," 29, 34.

12. Burnet, "Basis of allergic diseases," 30.

13. The influence on Burnet's immunological theories of his work on phage and research on the immune response to Staphylococcus toxin, the cause of the Bundaberg tragedy, is discussed in Sankaran, "The bacteriophage, its role in immunology"; Hooker, "Diphtheria, immunization and the Bundaberg tragedy"; and Hobbins, " 'Immunization is as popular as a death adder.' " On the influence of Burnet's ecological orientation, see Park, "Germs, hosts, and the origin of Frank Macfarlane Burnet's concept of 'self' and 'tolerance.' " More generally, see Sexton, *The Seeds of Time;* and Anderson, "Natural histories of infectious disease." Some attention should also be given to the influence of J. C. G. Ledingham, Burnet's Ph.D. supervisor at the University of London, and Christopher Andrewes, his close friend there, who had studied with Homer Swift at the Rockefeller Institute. See Andrewes, "Immunity in virus diseases." Burnet also seems to have read closely the work of Hans Zinsser (see Chapter 2 of this book). Alfred I. Tauber and his colleagues have noticed a similarity in the biological theories of Burnet and Elie Metchnikov, though there is no direct evidence of any influence. See Tauber, *The Immune Self;* Tauber and Podolsky, "Frank Macfarlane Burnet and the immune self"; and Crist and Tauber, "Selfhood, immunity, and the biological imagination."

14. Ehrlich, "On immunity with special reference to cell life." See also Chapter 2.

15. Pauling, "A theory of the structure and process of formation of antibodies." See also Breinl and Haurowitz, "Chemische Untersuchung des Prazipitätes aus Hämoglobin und Anti-Hämoglobin-Serum"; and Mudd, "A hypothetical mechanism of antibody formation."

16. Burnet, *Changing Patterns*, 192.

17. Burnet, "Antibody production as a special example of protein synthesis in vivo." Burnet favored publishing his more speculative theories in obscure Australian journals or monographs, so if they proved correct he could claim priority but if wrong he was saved embarrassment.

18. Burnet, *The Production of Antibodies.*

19. Burnet's equivocation reflected the general dispute in the 1940s between adaptationist and selectionist explanations of variation: see Creager, "Adaptation or selection?"; and Gould, "The hardening of the modern synthesis."

20. Burnet, "Antibody production in the light of recent genetic theory," 146. Burnet was interested in C. D. Darlington's conjectures about "plasmagenes" (Darlington, "Heredity, development and infection") and in other studies of cytoplasmic inheritance, especially the work of Tracy M. Sonneborn and Sol Spiegelman. See Sapp, *Beyond the Gene.*

21. Huxley, "The biological basis of individuality," 318, 310. In *Essays of a Biologist,* Huxley frequently refers to individuality (e.g., 108, 247).

22. While Burnet made a glancing reference to "self" (29) in *Biological Aspects of Infectious Disease*, the first edition of his *Production of Antibodies* (1941) significantly does not use the term. It is interesting to speculate that Burnet later chose to invoke self because he was aware that he was *not* devising a general theory of biological *individuality,* just an account of the adaptive immune response of vertebrates. Thomas Pradeu suggests that Burnet failed to provide a complete theory of biological individuality or identity, but perhaps that was not his intention (*The Limits of the Self*).

23. Burnet, *Changing Patterns,* 77. Burnet claimed, "Immunology has always seemed to me more a problem in philosophy than a practical science" ("The Darwinian approach to immunity," 17).

24. Whitehead, *Process and Reality,* esp. 57, 85, 108.

25. Whitehead, *Process and Reality,* 187, 241.

26. Agar, *A Contribution to the Theory of the Living Organism*, 8, 11. Burnet probably also had read Agar, "Whitehead's philosophy of organism." The influence of Gestalt psychology on Agar is evident, as well. Agar and Burnet knew each other through the University of Melbourne, the Wallaby and Boobook clubs, and the Eugenics Society, of which Agar was the first president. From 1944, Agar's office was across the road from Burnet's.

27. Gibson and Medawar, "The fate of skin homografts in man"; and Medawar, "The behavior and fate of skin autografts and skin homografts in rabbits," and "A second study of the behavior and fate of skin homografts in rabbits." Burnet first met Medawar in Oxford in 1946.

28. Burnet and Fenner, "Genetics and immunology," 316–17.

29. Owen, "Immunogenetic consequences of vascular anastomoses between bovine twins." Also, chimerism occurs frequently in twin chickens and more rarely in twin sheep. Owen, who had grown up on a Midwest dairy farm, left the University of Wisconsin for Caltech in 1947, moving from cattle barns to rodent labs, as he put it. See Crow, "A golden anniversary: cattle twins and immune tolerance." Another important influence was the finding that mice infected with the lymphocytic chorio-meningitis virus before birth later failed to produce antibody

against it, even though their tissues remained infected the rest of their lives (Traub, "Persistence of lymphocytic chorio-meningitis").

30. Burnet and Fenner, *The Production of Antibodies*, 2nd ed. According to Fenner, Burnet was "responsible for the all the interpretation and speculation" in the book (*Nature, Nurture and Chance*, 49). Burnet shared the 1960 Nobel Prize in Physiology or Medicine for his concept of immunological tolerance. Fenner became foundation professor of microbiology at the Australian National University, Canberra.

31. Burnet, "How antibodies are made," 76.

32. "Poor chap, we've got very much in common," Burnet wrote at age twenty-one after reading Barbellion's journal. "Bughunters, philosophizing, introspective, shy, a bit neurotic, seeing things a little clearer than most men and gaining little but unhappiness thereby" (General Diary, April 29, 1921, folder 6, series 2, Burnet papers).

33. Burnet, *Changing Patterns*; Sexton, *Seeds of Time*: Fenner, "Frank Macfarlane Burnet, 3 September 1899–31 August 1985"; and Neeraja Sankaran, "Frank Macfarlane Burnet and the Nature of Bacteriophage, 1924–1937," Ph.D. dissertation, Yale University, 2006.

34. Burnet, *Walter and Eliza Hall Institute*, 65. On the Hall Institute, see Vivianne de Vahl Davis, "A History of the Walter and Eliza Hall Institute of Medical Research, 1915–1978: An Examination of the Personalities, Politics, Finances, Social Relations and Scientific Organization of the Hall Institute," Ph.D. thesis, University of New South Wales, 1979; Nossal, *Diversity and Discovery*; and Charlesworth, Farrall, Stokes, and Turnbull, *Life Among the Scientists*. More generally on the development of postwar medical research in Australia, see Anderson, "The military spur to Australian medical research."

35. Burnet, "How antibodies are made," 78.

36. Burnet, *Enzyme, Antigen and Virus*, 164, 168, 174.

37. Burnet, *Enzyme, Antigen and Virus*, 166. In his memoirs, Burnet referred to this work as a "rather bad, over-ambitious book with an interpretation of antibody production which was already beginning to look unconvincing by the time I finalized the proofs" (*Changing Patterns*, 204).

38. F. M. Burnet to Linda Burnet, 8 March 1956, in folder 18, series 2, Burnet papers.

39. As Tauber notes, "Burnet sensed the potential power of information theory, but its application as a research program was highly problematic at this time" (*The Immune Self*, 164).

40. Quastler, "The measure of specificity," 41. Burnet owned probably the only copy of this book in Australia, and he referred to it frequently.

41. Wiener, *The Human Use of Human Beings*, 85, 90.

42. Haraway "The high cost of information in post–World War II evolutionary biology," 249. See also Kay, "Cybernetics, information, life," which briefly mentions Burnet and Medawar, along with Niels K. Jerne and Joshua Lederberg.

43. On other Cold War models of the self, see Lunbeck, "Identity and the real self in postwar American psychiatry"; and Cohen-Cole, "The creative American."

44. Medawar, "Burnet and immunological tolerance," 32.

45. Medawar, *Memoir of a Thinking Radish*, 112. Medawar also expressed enthusiasm for cybernetics; see his *Induction and Intuition in Scientific Thought.*

46. Anderson, Billingham, Lampkin, and Medawar, "The use of skin grafting."

47. Medawar quoted in Billingham, "Reminiscences of a transplanter," 81.

48. Billingham, Brent, and Medawar, "Actively acquired tolerance of foreign cells." For this work, Medawar shared with Burnet the 1960 Nobel Prize in Physiology or Medicine. See Brent, *A History of Transplantation Immunology*, and "Sir Peter Brian Medawar."

49. Gowans, "Peter Medawar: His life and work," 7.

50. Brent, "Tolerance and GVHD," 107. Burnet seems to have been oblivious. In 1956 he noted: "I seem to get on very well in that lab and Billingham and Brent were most helpful. . . . On my side I think I was able to give them some useful ideas for new experiments" (Burnet to Linda Burnet, 23 March 1956, folder 18, series 2, Burnet papers).

51. Söderqvist, *Science as Autobiography.*

52. Jerne, "The natural selection theory of antibody formation: Ten years later."

53. Jerne, "The natural-selection theory of antibody formation," 849. See also Söderqvist, "Darwinian overtones"; and Tauber and Podolsky, *The Generation of Diversity.* Jerne later developed the hemolytic plaque assay technique to detect single antibody producing cells, and the idiotypic network theory of immunological function (see Chapter 6). For his contributions to immunological theory he shared the 1984 Nobel Prize in Physiology or Medicine.

54. Jerne, "The natural-selection theory of antibody formation," 853.

55. Burnet, "The theories of antibody formation," 14.

56. Burnet, "The impact of ideas on immunology," 2.

57. Burnet, *Changing Patterns*, 204, 205.

58. Jerne, "Burnet and the clonal selection theory," 38.

59. Burnet, *Walter and Eliza Hall Institute*, Chapter 10; Whittingham and Mackay, "Design and functions of a department of clinical immunology"; and Anderson, "The reasoning of the strongest." Ian J. Wood, the first director of the CRU, describes its activities in his memoir, *Discovery and Healing in Peace and War.* See also Wood and Taft, *Diffuse Lesions of the Stomach.* Wood and Taft note that pernicious anemia, later recognized as having an autoimmune basis, results "from a primary atrophy of the gastric mucosa which may begin early in life and be determined by a genetic factor" (66).

60. Saint, King, Joske, and Finckh, "The course of infectious hepatitis with special reference to prognosis and the chronic stage," 119.

61. Jan G. Waldenström first described the condition in 1948, announcing it a few years later in an obscure publication and attributing it to a persisting virus in the liver ("Leber, Blutproteine und Nahrungseiweiss").

62. Kunkel, Ahrens, Eisenmenger, Bongiovanni, and Slater, "Extreme hypergammaglobulinemia in young women with liver disease"; and Zimmerman, Heller, and Hill, "Extreme hyperglobulinemia in subacute hepatic necrosis."

63. Joske and King, "The 'L.E. cell' phenomenon in active chronic viral hepatitis," 479. See also Mackay and Tait, "The history of autoimmune hepatitis"; and Mackay, "Historical reflections on autoimmune hepatitis."

64. D. C. Gajdusek to Mahtil [his mother] 12 May 1956, box 5, Gajdusek correspondence, MS c565, National Library of Medicine, Bethesda, MD. (Hereafter NLM.) In his journal he noted, "Above and over everything in the Institute hovers Sir Mac's envy of any fame in biology other than his own and his inability to tolerate originality and independent excellence in his staff" (22 February 1957, in *Journal 1955–1957*, 72).

65. For more on Gajdusek and his difficult relationship with Burnet, see Anderson, *The Collectors of Lost Souls*. Gajdusek was awarded the 1976 Nobel Prize in Physiology or Medicine for his discovery of the first human "slow virus," the putative cause of the neurological disease kuru.

66. A Melbourne medical graduate, Mackay developed tuberculosis after World War II and moved into a career in pathology. Studies in London with Sheila Sherlock and at the University of Washington, Seattle, turned his attention to liver disease and associated plasma protein abnormalities. See Mackay, "Burnet and autoimmunity"; and Anderson and Hunter, "'Wars have an overflow on everything': interview with Ian R. Mackay and Patricia Mackay." He had been seeking these diseases on the wards of the hospital. See also Mackay, Taft, and Cowling, "Lupoid hepatitis."

67. Gajdusek, "An 'auto-immune' reaction against human tissue antigens in certain chronic diseases," and "An 'autoimmune' reaction against human tissue antigens in certain acute and chronic diseases. I: Serological investigations"; and Mackay and Gajdusek, "An 'autoimmune' reaction against human tissue antigens in certain acute and chronic diseases. II: Clinical correlations." Despite the efforts of Mackay, chronic active hepatitis was not formally renamed autoimmune hepatitis until 1993: see Mackay, Weiden, and Hasker, "Autoimmune hepatitis."

68. D. C. Gajdusek to Barry Adels, December 12, 1956, box 1, Gajdusek correspondence, NLM.

69. D. C. Gajdusek to Joseph E. Smadel, December 29, 1956, box 31, Gajdusek correspondence, NLM. For more on Smadel, see Chapter 2. Believing that Gajdusek had tried to take all the credit, Mackay recalled, "Nobody felt they could enjoy a collaboration with that man" (Interviewed by Warwick Anderson, 13 January 2005, Melbourne). Despite this, Mackay and Gajdusek maintained their friendship.

70. Burnet, *Walter and Eliza Hall Institute*, 71.

71. Burnet, *Changing Patterns*, 230. See also Mackay, "The 'Burnet era' of immunology: origins and influence," and "Roots of and routes to autoimmunity."

72. Burnet, "Impact of ideas on immunology," 2. Burnet also met frequently during this period with R. A. Fisher, who was in the process of moving to Adelaide, Australia.

73. Burnet, "A modification of Jerne's theory of antibody production using the concept of clonal selection," 68, and *Clonal Selection Theory of Acquired Immunity.* See also Ada and Nossal, "The clonal-selection theory"; Nossal, "The coming of age of the clonal selection theory"; and Tauber and Podolsky, *Generation of Diversity.*

74. Burnet, "The mechanism of immunity," 61. Francis H. Crick had criticized Burnet for challenging, with his adaptive enzyme hypothesis, the central dogma of molecular biology ("On protein synthesis").

75. Brent, *History of Transplantation Immunology*, 13.

76. Gowans, "The effect of the continuous re-infusion of lymph and lymphocytes."

77. Simonsen, "The impact on the developing embryo and newborn animal of adult homologous cells," and "Graft-versus-host reactions."

78. Burnet, *The Clonal Selection Theory of Acquired Immunity*, 107. See also Gowans, McGregor, Cohen, and Ford, "Initiation of immune responses by small lymphocytes."

79. Burnet, *Clonal Selection Theory of Acquired Immunity*, 55.

80. Burnet, "A modification of Jerne's theory," 68. As Rudolf Virchow put it, "no matter how we twist and turn, we eventually come back to the cell" ("Cellular pathology," 81).

81. Burnet, "A Darwinian approach to immunity," 452.

82. Burnet, "Mechanism of immunity," 62.

83. Burnet, "Mechanism of immunity," 62, 64. Although congruent with Cold War thinking in its concern with security and secrecy, the word *forbidden* caused some uneasiness, even confusion, among immunologists, perhaps retarding acceptance of the theory.

84. Talmage, "Allergy and immunology." Talmage later moved to the University of Colorado at Boulder.

85. Talmage quoted in Cruse and Lewis, "David W. Talmage and the advent of the cell selection theory," 26.

86. Talmage in Cohn, Mitchison, Paul, Silverstein, Talmage, and Weigert, "Reflections on the clonal-selection theory," 826. See also Talmage, "Origin of the cell selection theories of antibody formation."

87. Burnet, *Changing Patterns*, 210. The "disproof" took the form of experimental results implying that lymphocytes produced multiple sorts of antibodies, not just one type—but this was later discredited.

88. Lederberg, "Ontogeny of the clonal selection theory," 179. See also Gurr and Viret, "From variation in genetic information to clonal deletion." Lederberg

was awarded the 1958 Nobel Prize in Physiology or Medicine for his discovery that bacteria can exchange genes.

89. Lederberg, "Genes and antibodies." Experimenters also showed it was possible to induce tolerance to large doses of antigens in neonatal life; see Weigle, "Immunological unresponsiveness."

90. Lederberg, "Ontogeny of the clonal selection theory," 179.

91. Nossal and Lederberg, "Antibody production by single cells." An Australian, Nossal succeeded Burnet as director of the Hall Institute. He discerned in Burnet a "singular blend of originality, intuition, naïve honesty, wisdom and depth of an almost spiritual kind, and daring conceptual breadth" ("Burnet and science—an appreciation," 315). See also Nossal, "One cell—one antibody"; and Marchalonis, "Burnet and Nossal."

92. Jerne, "The complete solution of immunology," 348. On assertions of closure in immunology, see Anderson, Rosenkrantz, and Jackson, "Toward an unnatural history of immunology"; and Silverstein, "The end is near! The phenomenon of the declaration of closure in a discipline."

93. Haurowitz, "The role of the antigen in antibody formation," 19.

94. Edelman and Benacerraf, "On structural and functional relations between antibodies and proteins," 1035.

95. Burnet, "Autoimmune disease: some general principles," 91, 92, 95. See also Burnet, "Autoimmune disease—experimental and clinical."

96. Burnet, *Changing Patterns*, 215.

97. F. M. Burnet, "Notes, Autoimmune Diseases, 1983," folder 136, series 6, Burnet papers.

98. Burnet, *Walter and Eliza Hall Institute*, 146. See also Mackay, "Autoimmunity since the 1957 clonal selection theory."

99. Burnet, "Autoimmune disease," 5.

100. Bielschowsky, Helyer, and Howie, "Spontaneous haemolytic anaemia in mice of the NZB/BL strain"; and Helyer and Howie, "Renal disease associated with positive lupus erythematosus tests in a crossbred strain of mice."

101. Holmes and Burnet, "The natural history of autoimmune disease in NZB mice"; and Burnet and Holmes, "Genetic investigations of autoimmune disease in mice."

102. Burnet, *Clonal Selection Theory of Acquired Immunity*, 122.

103. Burnet, "Darwinian approach to immunity."

104. Burnet, "The seeds of time," 306. As Rudolf Virchow would have it: "Life is different from processes in the rest of the world, and cannot be simply reduced to physical and chemical forces" ("One hundred years of general pathology," 207).

105. Burnet, "Men or molecules? A tilt at molecular biology," 37. See also Burnet, "Life's complexities: Misgivings about models."

106. Cohn in Cohn, Mitchison, Paul, Silverstein, Talmage and Weigert, "Reflections on the clonal-selection theory," 827.

107. Nossal, "Coming of age of the clonal selection theory," 48.

108. Jerne, "Summary: Waiting for the end," 591. See also Silverstein, *A History of Immunology*, Chapter 4.

Chapter 5. Doing Biographical Work

1. On illness "careers," see Goffman, *Asylums*; and Hughes, *The Sociological Eye*. On the expanded notion of illness "trajectories," see Strauss et al., *Chronic Illness and the Quality of Life*, 2nd ed. As Renée Fox put it in 1959, "illness is more than a biological condition; it is a social role with certain patterned characteristics and requirements" (*Experiment Perilous*, 115).

2. Corbin and Strauss, "Accompaniments of chronic illness." It is remarkable how often sociologists use autoimmune conditions to examine the effects of chronic illness, even if the mode of pathogenesis is not relevant to their analysis.

3. Kleinman, *The Illness Narratives*, 31.

4. Monks and Frankenberg, "Being ill and being me." See also Brooks and Matson, "Social-psychological adjustment to multiple sclerosis"; and Robinson, "Personal narratives, social careers and medical courses."

5. Aladjem, *The Sun Is My Enemy*. See Chapter 3 of this book.

6. Charmaz, "Loss of self," and "Struggling for a self." See also Bury, "Chronic illness as biographical disruption"; and Corbin and Strauss, "Accompaniments of chronic illness."

7. Rosenberg, "The tyranny of diagnosis."

8. Stewart and Sullivan, "Illness behavior and the sick role in chronic disease."

9. Chellingsworth, "Multiple sclerosis," 90.

10. Brown, "One man's experience with multiple sclerosis," 23.

11. Kinley, "MS: From shock to acceptance," 274, 275.

12. Brophy, *Baroque 'n' Roll*, 10, 11.

13. Brophy, *Baroque 'n' Roll*, 11, 1.

14. These case notes come from the private files of Ian R. Mackay.

15. See Chapter 3.

16. Holborow, Asherson, Johnson, Barnes, and Carmichael, "Antinuclear factor and other antibodies." It is still puzzling why, for example, antibodies to smooth muscle should indicate autoimmune liver disease. Conventional examination of tissues also can be used to aid diagnosis, such as identification of antibody-forming lymphocytes in the tissues affected in thyroiditis and Sjögren's disease.

17. Sever, "Major technological advances affecting clinical and diagnostic immunology." Sever concludes: "There is no doubt that the age of molecular immunology and molecular medicine is upon us" (3).

18. Brophy, *Baroque 'n' Roll*, 25, 25, 26.

19. Lowry, "One woman's experience with multiple sclerosis," 32.

20. Chellingsworth, "Multiple sclerosis," 93.

21. Kinley, "MS: from shock to acceptance," 275.

22. Rubinstein, *Take It and Leave It*, 16, 59, 97, 88, 107.

23. Howard, *Coming to Terms*, 7.

24. Howard, *Coming to Terms*, 22.

25. Howard, *Coming to Terms*, 77.

26. Felstiner, *Out of Joint*, 50, 55.

27. Felstiner, *Out of Joint*, 409, 419.

28. Felstiner, *Out of Joint*, 50. For other patient accounts of rheumatoid arthritis, see Markson, "Patient semeiology of a chronic disease"; Wiener, "The burden of rheumatoid arthritis"; and Avery, "Rip tide, swimming through life with rheumatoid arthritis."

29. Mackay observed in 1965, "This man is a mine of autoimmune disease." He wrote to the patient's family doctor: "His life prognosis should be reasonably good—at least beyond five years, in my opinion."

30. For example, see Whittingham, Mackay, and Kiss, "An interplay of genetic and environmental factors."

31. Kleinman, *Illness Narratives*, 8.

32. Charmaz, "Loss of self," and "Struggling for a self."

33. Corbin and Strauss, "Accompaniments of chronic illness."

34. Hench, "Potential reversibility of rheumatoid arthritis." A medical writer received a tip and excitedly reported the findings the next day: William L. Lawrence, "Aid in rheumatoid arthritis is promised by new hormone," *New York Times* (21 April 1949): 1, 4.

35. Kendall, *Cortisone: Memoirs of a Hormone Hunter;* Polley and Slocumb, "Behind the scenes with cortisone and ACTH"; Glyn, "The discovery and early use of cortisone"; and Rasmussen, "Steroids in arms." Within a few years this sort of experiment would be forbidden: see Lederer, *Subjected to Science*. Hench and Kendall (with Tadeus Reichstein) shared the 1950 Nobel Prize in Physiology or Medicine.

36. Polley and Slocumb, "Behind the scenes with cortisone," 474.

37. Philip S. Hench—Nobel Lecture, Nobelprize.org, 8 March 2012. http://www.nobelprize.org/nobel_prizes/medicine/laureates/1950/hench-lecture.html.

38. Freyberg, Traeger, Patterson, et al., "Problems of prolonged cortisone treatment"; and Maddocks, "The effects of prolonged corticosteroid therapy."

39. A reference to Cushing's syndrome, caused by benign tumors producing high levels of cortisol and adrenocorticotropic hormone (ACTH).

40. Marks, "Cortisone, 1949"; and Cantor, "Cortisone and the politics of drama." More generally, see Liebenau, *Medical Science and Medical Industry;* Swann, *Academic Scientists and the Pharmaceutical Industry;* and Marks, *The Progress of Experiment.*

41. Hettenyi and Karsch, "Cortisone therapy"; and Slater, "Industry and academy." On the pharmaceutical exploitation of *barbasco*, see Soto Laveaga, *Jungle Laboratories.*

42. Mackay and Burnet, *Autoimmune Diseases*, 262.

43. Miller, Newell, and Ridley, "Multiple sclerosis."

44. Mackay and Burnet, *Autoimmune Diseases*, 249.

45. Gilman and Phillips, "The biological actions and therapeutic applications." "Incidents" refers in part to the "Bari incident" in 1943, when a Luftwaffe bombing in that Italian port of an Allied ship containing mustard gas killed almost one thousand people over the following months.

46. On their research program, see Hitchings, Elion, Falco, Russell, and Werff, "Studies on analogs of purines and pyrimidines." See also Elion, "Historical background of 6-mercaptopurine." Elion and Hitchings received the 1988 Nobel Prize in Physiology or Medicine for their discoveries of important principles for drug treatment.

47. Farber, Diamond, Mercer, Sylvester, and Wolff, "Temporary remissions in acute leukemia in children." For a vivid account of these developments, see Mukherjee, *The Emperor of All Maladies*.

48. Allison, "Immunosuppressive drugs." See Schwartz and Dameshek, "Drug-induced immunological tolerance." There was logic in using anticancer agents to suppress autoimmune disease, since, according to Mackay and Burnet, clonal proliferation functioned as "a conditioned neoplasm" (*Autoimmune Diseases*, 246). Of course, azathioprine and methotrexate also cause unwanted side effects. Both can induce nausea, fatigue, hair loss, and increased susceptibility to infection, and methotrexate might further give rise to headaches and liver damage, suppress the formation of blood cells, and lead, rarely, to lymphoma or leukemia.

49. Macfarlane Burnet, *Self and Not-Self*.

50. Flannery O'Connor to Caroline Gordon Tait, 10 December 1957, in O'Connor, *The Habit of Being*, 257. In *Experiment Perilous*, Renée Fox reported on the introduction of ACTH (adrenocorticotropic hormone) and cortisone on a ward of Boston's Peter Brent Brigham Hospital in the early 1950s.

51. Flannery O'Connor to A [Betty Hester], 24 December 1960, in O'Connor, *Habit of Being*, 423.

52. These case notes come from the private files of Ian R. Mackay. See Mackay, Wieden, and Ungar, "Treatment of chronic active hepatitis." For a description of the miracles worked by cortisone on rheumatoid arthritis, see Fisher, *Journeying On*.

53. Mackay, "Roots of and routes to autoimmunity."

54. Lancereaux, "Le diabète maigre."

55. Draper, Dupertuis and Caughey, "The differentiation by constitutional methods between pancreatic diabetes and diabetes of pituitary origin." On constitutional medicine and somatotypes, see Tracy, "An evolving science of man."

56. Gale, "The discovery of type I diabetes." On diabetes before and after insulin, see Bliss, *The Discovery of Insulin*; and Feudtner, *Bittersweet*.

57. Gundersen, "Is diabetes of infectious origin?"

58. Gepts, "Pathologic anatomy of the pancreas."

59. Renold, Soeldner, and Steinke, "Immunological studies with homologous and heterologous pancreatic insulin"; and Lacy and Wright, "Allergic interstitial pancreatitis."

60. Botazzo, Florin-Christensen, and Doniach, "Islet cell antibodies in diabetes mellitus."

61. These quotations are from Diabetes Stories, Personal Tales of Diabetes Through the Decades at http://www.diabetes-stories.com (accessed 1 May 2012).

62. Corris, *Sweet and Sour*, 27, 78.

63. Fred Hollows quoted in Corris, *Sweet and Sour*, 118.

64. Anderson, *The Collectors of Lost Souls*.

65. Ziegler and Stites, "Hypothesis: AIDS is an autoimmune disease." See also Löwy, "Immunology and AIDS."

66. There does appear to be an intriguing association between HIV infection and the development of autoimmune disease; see Zandman-Goddard and Schoenfeld, "HIV and autoimmunity."

67. See Sturdy, "Looking for trouble," esp. 744.

68. Mackay wondered if a study of steroids and azathioprine "that included randomly allocated placebo control patients was ethically problematic" ("Historical reflections on autoimmune hepatitis," 3295).

69. Mackay, "Burnet and autoimmunity."

70. On self-help groups, see Gussow and Tracy, "The role of self-help clubs in adaptation to chronic illness"; and Maines, "The social arrangements of diabetic self-help groups."

71. Rabinow, "Artificiality and enlightenment."

72. Felstiner, *Out of Joint*, 164, 581, 599.

Chapter 6. Reframing Self

1. Landsteiner and Chase, "Studies in the sensitization of animals."

2. Landsteiner and Chase, "Experiments on transfer of cutaneous sensitivity."

3. Chase, "Irreverent recollections, from Cooke and Coca, 1928–1978," 309.

4. See Landsteiner's notes on Chase in folder 6, box 1, RG 450 L239, Karl Landsteiner Papers, Rockefeller University Archives, Rockefeller Archive Center, New York. (Hereafter, RUA.)

5. Chase, "The cellular transfer of cutaneous sensitivity to tuberculin," and "The allergic state." Later studies of graft rejection, graft-versus-host disease, and experimental allergic encephalomyelitis indicated transfer of immunological reactivity exclusively through cells and not antibodies.

6. Merrill W. Chase, in Reports to the Corporation and to the Board of Scientific Directors 38 (1949–1950): 44, folder 1, box 25, RG 439, RUA. Chase also had read a neglected treatise, Murphy, *The Lymphocyte in Resistance to Tissue Grafting, Malignant Disease, and Tuberculous Infection*. Even into the 1960s, some immunologists countered the cellular emphasis by postulating cell-bound antibodies: see Amos

and Koprowski, ed., *Cell-Bound Antibodies;* and Silverstein, "Whatever happened to cell-bound antibodies?"

7. See Chapter 4.

8. Doherty, "Challenged by complexity," 6.

9. Miller, "The discovery of thymus function," 7, 8. Born in Nice, France, and brought up in Shanghai (where his father managed a bank) and Sydney, Miller attended high school and medical school with Gustav Nossal, another émigré. "Miller" was a convenient version of Meunier. See also Miller, "Discovering the origins of immunological competence."

10. Miller, "The discovery of thymus function," 9.

11. Miller, "The discovery of the immunological function of the thymus," *Immunology Today* 12 (1991): 42–45, 43.

12. Miller, "Discovery of the immunological function of the thymus," 43. See Miller, "Immunological function of the thymus."

13. Miller, "Discovery of thymus function," 9. See Miller, "Effect of neonatal thymectomy on the immunological responsiveness of the mouse." Since Miller pronounced "thymus" as "sinus," his early lectures caused considerable confusion (Jacques Miller, interviewed by Warwick Anderson, 3 February 2014, Melbourne). Peter Medawar was skeptical: "We shall come to regard the presence of lymphocytes in the thymus," he wrote, "as an evolutionary accident of no very great significance" (quoted in *The Immunologically Competent Cell*, eds. Wolstenholme and Knight, 70). In 1966, Miller became a researcher at the Walter and Eliza Hall Institute, Melbourne, Australia.

14. Maclean, Zak, Varco and Good, "Thymic tumor and acquired agammaglobulinemia."

15. Good, Peterson, Martinez, Sutherland, Kellum, and Finstad, "The thymus in immunobiology," 75. Good initially equivocated on whether his thymectomies had any significant effect on the immune response, and certainly his results in rabbits were ambiguous. As Jacques Miller later reflected, Good's "thymectomized rabbits were not 'immunological cripples'—that's just embellishment . . . the work was clearly done in a hurry" (interviewed by Warwick Anderson, 3 February 2014, Melbourne).

16. Good et al., "The thymus in immunobiology," 76. Confusion over priority is probably responsible for no award of a Nobel Prize for the discovery of the function of the thymus. See Good, Dalmasso, Martinez, Archer, Pierce, and Papermaster, "The role of the thymus in development of immunologic capacity in rabbits and mice." Good became director of the Sloan-Kettering Institute for Cancer Research (1973–82).

17. Glick, "The saga of the bursa of Fabricius." Hieronymus Fabricius was a Paduan anatomist who taught William Harvey.

18. Chang, Glick, and Winter, "The significance of the bursa of Fabricius"; and Glick, Chang, and Jaap, "The bursa of Fabricius and antibody production."

19. Glick, "The saga of the bursa of Fabricius," 191. See also Ribatti, Crivellato, and Vacca, "The contribution of Bruce Glick to the definition of the role played by the bursa of Fabricius." Glick later taught ornithology at Mississippi State and Clemson.

20. Good et al., "The thymus in immunobiology," 75. See Papermaster, Friedman, and Good, "Relationships of the bursa of Fabricius to immunological responsiveness."

21. Warner, Szenberg, and Burnet, "The immunological role of different lymphoid organs in the chicken"; and Warner and Szenberg, "Dissociation of immunological responsiveness in fowls," and "Effect of neonatal thymectomy on the immune response of the chicken."

22. Claman, Chaperon, and Triplett, "Thymus-marrow cell combinations."

23. Miller and Mitchell, "Cell to cell interaction in the immune response. I: Hemolysin-forming cells in neonatally thymectomized mice reconstituted with thymus or thoracic duct lymphocytes"; and Mitchell and Miller, "Cell to cell interaction in the immune response. II: The source of hemolysin-forming cells in irradiated mice given bone marrow and thymus or thoracic duct lymphocytes," and "Immunological activity of the thymus and thoracic-duct lymphocytes."

24. Miller, "Discovery of thymus function," 12. Miller writes that Burnet expressed doubt about the value of such "biological monstrosities" as thymectomized, irradiated, bone-marrow reconstituted mice (12). See also Miller, "Antigen-specific clone selection *in vivo*." On the discovery in the late 1960s of the cytokines that convey nonspecific messages between cells, see Waksman and Oppenheim, "The contribution of the cytokine concept to immunology."

25. Roitt, Greaves, Torrigiani, Bostoff, and Playfair, "The cellular basis of immunological responses."

26. Mitchell, "Selection, memory and selective memories," 26.

27. Roitt et al., "Cellular basis of immunological responses." In arguing that T cells were the repositories of immunological memory, Roitt was following Miller and Sprent, "Cell to cell interaction in the immune response."

28. Gershon and Kondo, "Cell interactions in the induction of tolerance," and "Infectious immunological tolerance"; and Gershon, Cohen, Hencin, and Liebhaber, "Suppressor T cells." See also Agyris, "Adoptive tolerance."

29. Cohn, "Commentary," 195. See also Möller, "Do suppressor T cells exist?" On the controversy, see Sercarz, Oki, and Gammon, "Central versus peripheral tolerance"; and Keating and Cambrosio, "Helpers and suppressors."

30. Sakaguchi, "Naturally arising CD4 regulatory T cells for immunologic self tolerance and negative control of immune responses."

31. Chatenoud, Salomon, and Bluestone, "Suppressor T cells."

32. Steinman and Cohn, "Identification of a novel cell type in peripheral lymphoid organs of mice"; and Banchereau and Steinman, "Dendritic cells and the con-

trol of immunity." Ralph M. Steinman shared the 2011 Nobel Prize in Physiology or Medicine for his discovery of dendritic cells.

33. Rose, "Reflections on tolerance, self-tolerance and Felix Milgrom," 1460. Charles Janeway, Jr., called for the "sociology of lymphocytes" ("Approaching the asymptote?" 12).

34. Rose, "Reflections on tolerance," 1462.

35. Avrameus, "Natural autoantibodies: from 'horror autotoxicus' to 'gnothi seauton,'" and "Natural autoantibodies: self-recognition and physiological autoimmunity."

36. Dausset, "The HLA adventure," 14.

37. Dausset, "The major histocompatibility complex in man," 628.

38. Klein, "Seeds of time."

39. Gorer, "The genetic and antigenic basis of tumour transplantation," and "The antigenic basis of tumour transplantation."

40. Snell, ed., *Biology of the Laboratory Mouse*. See also Gaudillière, "Mapping as technology." Snell had worked at the Jackson Laboratory with Clarence C. Little, an expert on tumor transplantation. See Snell, "A geneticist's recollections of early transplantation studies."

41. Gorer, Lyman, and Snell, "Studies on the genetic and antigenic basis of tumour transplantation." Snell shared the 1980 Nobel Prize in Physiology or Medicine; Gorer died young, in 1961.

42. Brent, *A History of Transplantation Immunology*, 135.

43. Soulillou, "An interview with Jean Dausset." See also Jean Dausset, "The HLA adventure," and *Clin d'Oeil à la Vie*; and Dausset and Rapaport, "The HLA story."

44. Dausset and Nenna, "Présence d'une leuco-agglutinine dans le sérum d'un cas d'agranulocytose chronique." See also Dausset, "Leuco-agglutinins. IV: Leuco-agglutinins and blood transfusion."

45. Dausset, "Iso-leuco-anticorps."

46. Brent, *A History of Transplantation Immunology*, 141. See also Payne, "Early history of HLA"; and Rood and Leeuwen, "The HLA story as seen from Leiden."

47. Löwy, "The impact of medical practice on biomedical research."

48. Dausset and Rapaport, "The role of blood group antigens in human histocompatibility."

49. Dausset founded the Centre d'Études du Polymorphisme Humain, which collaborated in the human genome diversity program. He shared the 1980 Nobel Prize in Physiology or Medicine with George Snell.

50. Benacerraf, "Reminiscences," and *From Caracas to Stockholm*. By marriage, Benacerraf was Jacques Monod's nephew.

51. Levine, Ojeda, and Benacerraf, "Studies on artificial antigens"; and Green and Benacerraf, "Genetic control of immune responsiveness." See also Benacerraf, "Role of MHC gene products in immune regulation."

52. McDevitt and Chinitz, "Genetic control of the antibody response"; and McDevitt and Bodmer, "Histocompatibility antigens, immune responsiveness and susceptibility to disease." See also McDevitt, "Discovering the role of the major histocompatibility complex in the immune response." Hal Holman (see Chapter 3) recruited McDevitt to Stanford.

53. Benacerraf and McDevitt, "Histocompatibility-linked immune response genes," 275, 277. Benacerraf shared the 1980 Nobel Prize for Physiology or Medicine. According to Benacerraf, McDevitt "had a very keen mind, an enthusiastic personality, and a willingness to become familiar with the newest techniques of transplanation genetics" ("Reminiscences," 20).

54. Benacerraf and McDevitt initially proposed the T cell receptor as the product of the immune response genes but soon abandoned the idea.

55. Grumet, Conkel, Bodmer, Bodmer, and McDevitt, "Histocompatibility (HLA) antigens associated with systemic lupus erythematosus."

56. Vladutiu and Rose, "Autoimmune murine thyroiditis relation to histocompatibility (H-2) type." See also Noel R. Rose, "Autoimmunity: A personal memoir," and "Thyroid autoimmunity: A voyage of discovery."

57. Naito, Namerow, Mickey, and Terasaki, "Multiple sclerosis: Association with HL-A3"; and Mackay and Morris, "Association of autoimmune active chronic hepatitis with HL-A1, 8."

58. Nerup, Platz, Andersen, Christy, et al., "HL-A antigens and diabetes mellitus." See also Nepom and Erlich, "MHC class II molecules and autoimmunity."

59. McDevitt and Bodmer, "HL-A, immune-response genes, and disease," 1274.

60. Having studied medicine in Basel, Paris and Berlin, Zinkernagel, a Swiss national, was a postdoctoral fellow in Canberra. He later returned to Switzerland, as professor of microbiology at the University of Zurich. See Zinkernagel, "About the discovery of MHC-restricted T cell recognition." A Queensland graduate in veterinary medicine, Doherty had studied the inflammatory response to louping ill encephalitis at the Moredun Research Institute, Edinburgh, before coming to Canberra to learn more about cellular immunity, in particular the role of T cells. See Doherty, "Living in the Burnet lineage," and "Challenged by complexity."

61. Zinkernagel and Doherty, "Restriction of *in vitro* T cell-mediated cytotoxicity in lymphocytic chorio-meningitis within a syngeneic or semiallogeneic system."

62. Zinkernagel and Doherty, "Immunological surveillance against altered self components by sensitized T lymphocytes," and "MHC-restricted cytotoxic T cells." Zinkernagel and Doherty shared the 1996 Nobel Prize in Physiology or Medicine.

63. Doherty, *The Beginner's Guide to Winning the Nobel Prize*, 125. Doherty recalled that at Stanford he "gave this talk about altered self, and people later said, 'You know, we realized that what you were saying was important but we didn't really get it.'

Because what we [Zinkernagel and Doherty] were doing was turning transplantation on its head. We were saying that the MHC molecule is a marker of 'self' rather than 'foreignness'" (interviewed by Warwick Anderson, 17 March 2014, Melbourne).

64. Benacerraf, "Role of MHC gene products," 1236.

65. See Anne Marie Moulin, "La métaphore du soi et le tabou de l'autoimmunité"; Tauber, "The immunological self: a centenary perspective"; and Löwy, "The immunological construction of the self."

66. Cohn, "Immunology: What are the rules of the game?" 11, 12. See also Langman and Cohn, "A minimal model for the self-nonself discrimination."

67. Tauber and Podolsky, *The Generation of Diversity*; and Chadarevian and Kamminga, eds., *Molecularizing Biology and Medicine.*

68. Anne Marie Moulin argues the notion of the immune *system* emerged in the late 1960s and took off in subsequent decades; see "The immune system: a key concept in the history of immunology."

69. Jerne, "Toward a network theory of the immune system," 383, 387. An admirer of Søren Kierkegaard, Jerne would have been aware of the philosopher's dictum that "the self is the conscious synthesis of infinitude and finitude that relates itself to itself" (*Sickness unto Death*, 29).

70. Jerne, "Idiotypic networks and other preconceived ideas," 19.

71. Cohn, "The wisdom of hindsight," 39, 40.

72. Cohn, "The immune system," 25.

73. Vaz and Varela, "Self and nonsense: An organism-centered approach to immunology," 231. See also Coutinho, Forni, Holmberg, Ivars, and Vaz, "From an antigen-centered, clonal perspective of immune responses to an organism-centered, network perspective."

74. Vaz and Varela, "Self and nonsense," 238 (original emphasis).

75. Vaz and Varela, "Self and nonsense," 257; and Varela and Anspach, "The body thinks: The immune system and the process of somatic individuation," 282.

76. Varela, "Organism: a meshwork of selfless selves."

77. Cohen, "Kadishman's tree, Escher's angels, and the immunological homunculus," 11. See also Cohen, "The cognitive principle challenges clonal selection," and *Tending Adam's Garden*. For an earlier use of the term "immunological homunculus," see Whittingham and Mackay, "The 'pemphigus' antibody and the immunopathies affecting the thymus," 5.

78. See Agamben, *Homo Sacer;* and Foucault, *Discipline and Punish*, and "*Society Must be Defended.*" Tauber interprets this as a shift from a modern to a postmodern self; see "Immunology and the enigma of selfhood."

79. Cohen, "Kadishman's tree," 12, 16.

80. Cohen, "Biomarkers, self-antigens and the immunological homunculus," 248.

81. Janeway, "Approaching the asymptote."

82. Janeway, "The immune system evolved to discriminate infectious nonself from non-infectious self." See also Janeway, "A trip through my life with an

immunological theme." It was determined later that dendritic cells also could recognize self-antigens.

83. Matzinger, "Tolerance, danger, and the extended family," 991. Matzinger was an immunologist at the National Institute of Allergic and Infectious Diseases, Bethesda, MD.

84. Matzinger, "Tolerance, danger, and the extended family," 1038. For models of the immune response that require more than one activating signal, see Forsdyke, "The liquid scintillation counter as an analogy for the distinction between 'self' and 'notself' in immunological systems," and "Immunology (1955–1975): The natural selection theory, the two signal hypothesis and positive repertoire selection"; Bretscher and Cohn, "A theory of self-nonself discrimination"; and Lafferty and Cunningham, "A new analysis of allogenic interactions."

85. Matzinger, "The danger model: a renewed sense of self," 304. See also Matzinger, "The danger model in its historical context," and "The evolution of the danger theory." Surprisingly, there is little discussion in the immunological literature of the lymphocyte response to ordinary tissue damage, the result, for example, of myocardial infarction, ischemic stroke, or crush injuries.

86. Vance, "A Copernican revolution? Doubts about the danger theory," 1725, 1727, 1726. For other criticisms, see Silverstein and Rose, "On the mystique of the immunological self," and "There is only one immune system! The view from immunopathology"; Pradeu and Cooper, "The danger theory: 20 years later"; and Pradeu, *The Limits of the Self.*

87. Zinkernagel, "Uncertainties—discrepancies in immunology," 121.

88. Valéry, *Cahiers*, vol. 2, 1356.

89. Temkin, "The scientific approach to disease"; Canguilhem *The Normal and the Pathological;* Starobinski, *Action and Reaction;* and Moulin, "The dilemma of medical causality and the issue of biological individuality," and "Multiple splendor."

90. Doherty, "Endings and beginnings," 1. Doherty has said, "I can never understand why, if you have the network, the whole thing didn't reify into a lattice and become totally constipated" (interviewed by Warwick Anderson, 17 March 2014, Melbourne).

91. Jerne, "The somatic generation of immune recognition."

92. Zinkernagel, "Uncertainties—discrepancies in immunology," 103.

93. Coutinho and Kazatchkine, "Autoimmunity today."

94. Rose and Mackay, "Molecular mimicry"; and Mackay, "The etiopathogenesis of autoimmunity." See also Mackay, "Autoimmunity: Paradigms of Burnet and complexities of today," and "Autoimmunity since the 1957 clonal selection theory."

Afterword: Becoming Autoimmune, or Being Not

1. See Tauber, "The molecularization of immunology"; and Strasser and Fantini, "Molecular diseases and diseased molecules."

2. Kang and Craft, "Systemic lupus erythematosus: Immunologic features"; and Yidirim-Toruner and Diamond, "Current and novel therapeutics in treatment of SLE."

3. See Chapters 3 and 5.

4. Mackie, Quinn, and Emery, "Rheumatoid arthritis."

5. Barbellion, *The Journal of a Disappointed Man*, 5 July 1917, 292. See Chapter 1.

6. Brass, Weiner, and Hafler, "Multiple sclerosis."

7. Bach, "Type 1 diabetes."

8. The influence of gut microorganisms, the intestinal microbiome, on the maturation of the immune system is exciting much scientific interest. While some pathogens are putatively associated with the development of autoimmune disease, it seems that intestinal flora might act either to inhibit or stimulate autoimmunity. This may explain the apparent association of the increase of autoimmune disease and allergy with decreased infection and improved hygiene. Another intriguing line of inquiry is the relation of apoptosis or programmed cell death to autoimmunity. There is some evidence that defects in apoptosis and the clearance of necrotic cells might induce auto-reactivity.

9. Burnet, "The Darwinian approach to immunity," 17. Foucault noted in his study of the Paris hospitals in the early nineteenth century, "In this culture medical thought is fully engaged with the philosophical status of man" (*The Birth of the Clinic*, 245).

10. Whitehead, *Science and the Modern World*, 17.

11. Haraway, "The biopolitics of postmodern bodies," 204, 211. One might argue that Ludwik Fleck, even in the 1930s, was creating a sort of immunological philosophy of science; see Chapter 2.

12. Martin, "The end of the body?" 126.

13. Derrida, *Rogues*, 124.

14. Derrida, "Faith and knowledge," 73n.

15. Wittgenstein, "Philosophy," 161, and *Philosophical Investigations*, 255. We thank Huw Price for drawing our attention to this. Of course, it is likely that Wittgenstein would have regarded immunological philosophy as a form of bewitchment rather than therapy.

16. Löwy, "The strength of loose concepts," 89; and Tauber, "The immune self," 136.

17. Derrida, *Rogues*, 109.

18. Burnet, *The Integrity of the Body*, v–vi.

19. Locke, *An Essay Concerning Human Understanding*. Locke emphasized the capacity of the self for detachment and independence, that is, its relational qualities.

20. Vidal, "Brains, bodies, selves, and science," 939. See also Heller, Sosna, and Wellburn, eds., *Reconstructing Individualism;* Taylor, *Sources of the Self;* Martin and Barresi, *The Rise and Fall of Soul and Self;* and Nikolas Rose, "How should one

do the history of the self?" There is an alternative discussion of the self in twentieth-century social science: see, for example, Mauss, "A category of the human mind"; and Burkitt, "The shifting concept of the self."

21. As Emily Martin points out, the emergence of the acquired immunodeficiency syndrome (AIDS) in the 1980s brought immunology to prominence in philosophical discussions of embodiment and self.

22. Morse, "Something in the air." Sloterdijk argues that human culture depends on immunizing spheres, which modernization and globalization have disrupted or destroyed—giving rise to the need for pluralistic co-immunizing strategies. See Sloterdijk, *Sphären III;* and Sloterdijk and Heinrichs, *Die Sonne und der Tod.*

23. Haraway, "Biopolitics of postmodern bodies," 205, 207.

24. Haraway, "Biopolitics of postmodern bodies," 211, 212.

25. Haraway, "Biopolitics of postmodern bodies," 218.

26. Martin, "Toward an anthropology of immunology," 417.

27. Martin, "End of the body," 123, 129.

28. Martin, *Flexible Bodies,* 111, 93.

29. Napier, *The Age of Immunology,* 3.

30. Napier, "Non self help," 125.

31. Napier, "Non self help," 130, 132, 133 (original emphasis).

32. Napier, "Non self help," 132, 134.

33. Derrida, *Specters of Marx,* 177 (original emphasis). By the late 1980s, Derrida had come to appreciate the challenge of AIDS to the metaphysics of the subject. "AIDS, an event that one could call historial in the epoch of subjectivity," he said to Jean-Luc Nancy in 1991, "if we still gave credence to historiality, to epochality, and to subjectivity" ("'Eating well,'" 285). Derrida was an assistant to Georges Canguilhem in the early 1960s. As Rheinberger insists, "we have to situate Derrida's work with respect to science in a more general sense" ("Translating Derrida," 180).

34. Derrida, *Specters of Marx,* 221.

35. Mitchell, "Picturing terror," 281, 286.

36. Naas, "'One nation . . . indivisible,'" 17.

37. Derrida, "Faith and knowledge," 73n. He repeats this definition in Derrida, "Autoimmunity," 94.

38. For example, Ziegler and Stites, "Hypothesis: AIDS is an autoimmune disease." By the early 1990s, AIDS was commonly attributed solely to infection by the human immunodeficiency virus (HIV). See also Löwy, "Immunology and AIDS"; and Chapter 5.

39. Derrida, "Faith and knowledge," 73n, 47 (original emphasis).

40. Freud, *Beyond the Pleasure Principle.*

41. Derrida, "Faith and knowledge," 51. This remark echoes Hegel on fever; see Chapter 1.

42. Esposito in Campbell, "Interview: Roberto Esposito," 53, 54.

43. Esposito, *Immunitas*, 159, 17, 18. Esposito is influenced by Jean-Luc Nancy's analysis of the relations of the body to technical attachments; see, in particular, Nancy's account of his own heart transplant, in which "my self becomes my intruder" (*Corpus*, 168).

44. Esposito, *Immunitas*, 141.

45. Esposito, *Immunitas*, 174. See also Campbell, "Bios, immunity, life."

46. Taylor, "The betrayal of the body," 246.

47. Taylor, "Betrayal of the body," 251.

48. Taylor, "Betrayal of the body," 1.

49. Taylor, "Betrayal of the body," 253, 255 (original emphasis).

50. Andrews, "Autoimmune illness as a death-drive," 189.

51. Andrews, "Autoimmune illness as a death-drive," 189, 190.

52. Andrews, "Autoimmune illness as a death-drive," 190.

53. Andrews, "Autoimmune illness as a death-drive," 193 (original emphasis).

54. Derrida, *Rogues*, 150–51.

55. Manguso, *The Two Kinds of Decay*, 13.

56. Manguso, *Two Kinds of Decay*, 19. Her formal diagnosis was chronic inflammatory demyelinating polyneuropathy.

57. Manguso, *Two Kinds of Decay*, 131, 180. On feminist reassertions of embodied experience, see Grosz, *Volatile Bodies;* and Braidotti, *Nomadic Subjects*.

58. Clendinnen, *Tiger's Eye*, 1.

59. Clendinnen, *Tiger's Eye*, 9, 10.

60. Clendinnen, *Tiger's Eye*, 15, 18.

61. Clendinnen, *Tiger's Eye*, 288, 286.

62. Clendinnen, *Tiger's Eye*, 288–89.

BIBLIOGRAPHY

Interviews

Conducted by Warwick Anderson

Peter C. Doherty, March 17, 2014, Melbourne, Victoria.
Frank Fenner. July 29, 2002, Canberra, Australian Capital Territory.
Halstead R. Holman, November 12, 2012, Stanford, California.
Ian R. Mackay, January 13, 2005, Melbourne, Victoria.
Jacques F. P. Miller, February 3, 2014, Melbourne, Victoria.
Gustav J. V. Nossal, March 17, 2014, Melbourne, Victoria.
Ivan M. Roitt, June 3, 2013, Golders Green, London.
Noel R. Rose, April 30, 2012, Baltimore, Maryland.

Archives

Australia

Frank Macfarlane Burnet Papers, 1989.0034, University of Melbourne Archives, Melbourne, Victoria.
Private papers of Ian R. Mackay, Melbourne, Victoria.

United States of America

D. Carleton Gajdusek Correspondence, MS C565, National Library of Medicine, Bethesda, Maryland.
Robert A. Good Papers, HMS c317, Countway Medical Archives, Harvard University.
Henry G. Kunkel Papers, Rockefeller University Archives, Rockefeller Archive Center, New York.
Karl Landsteiner Papers, Rockefeller University Archives, Rockefeller Archive Center, New York.
Reports to the Corporation and to the Board of Scientific Directors, Rockefeller University Archives, Rockefeller Archive Center, New York (cited as Rockefeller Institute Scientific Reports).
Hans Zinsser Papers, HMS c73, Countway Medical Archives, Harvard University.

Published Primary Sources

Adler, Alfred. "On the Neurotic Disposition [1909]." In *The Collected Works of Alfred Adler, Vol. 2: Journal Articles, 1898–1909; A Study of Organ Inferiority,*

1907: The Mind Body Connection, Social Activism, and Sexuality, edited by Henry T. Stein, translated by Gerald L. Liebenau. Bellingham, WA: Alfred Adler Institute, 2002.

Agar, W. E. *A Contribution to the Theory of the Living Organism.* Melbourne: Melbourne University Press with Oxford University Press, 1943.

———. *Cytology: With Special Reference to the Metazoan Nucleus.* London: Macmillan, 1920.

———. "Whitehead's Philosophy of Organism: An Introduction for Biologists." *Quarterly Review of Biology* 11 (1936): 16–34.

Agyris, B. F. "Adoptive Tolerance: Transfer of the Tolerant State." *Journal of Immunology* 90 (1963): 29–34.

Aladjem, Henrietta. *The Sun Is My Enemy: One Woman's Victory over a Mysterious and Dreaded Disease.* Englewood Cliffs, NJ: Prentice-Hall, 1972.

Aladjem, Henrietta, and Peter H. Schur. *In Search of the Sun: A Woman's Courageous Victory over Lupus.* New York: Charles Scribner's Sons, 1988.

Allison, Anthony C. "Immunosuppressive Drugs: The First Fifty Years and a Glance Forward." *Immunopharmacology* 47 (2000): 63–83.

Amos, Bernard, and Hilary Koprowski, eds. *Cell-Bound Antibodies.* Philadelphia: Wistar Institute, 1963.

Anderson, D., R. E. Billingham, G. H. Lampkin, and P. B. Medawar. "The Use of Skin Grafting to Distinguish between Monozygotic and Dizygotic Twins in Cattle." *Heredity* 5 (1951): 379–97.

Anderson, McCall. "Progress of Medical Science." *Medical Record* (27 August 1887): 248.

Andral, [Gabriel]. "On the Physical Alterations of the Blood and Animal Fluids in Disease." *Provincial Medical and Surgical Journal* 2 (1841): 419–21.

Andrewes, C. H. "Immunity in Virus Diseases." *Lancet* 217 (1931): 989–92, 1046.

Andrews, Alice. "Autoimmune Illness as a Death-Drive: An Autobiography of Defence." *Mosaic* 44 (2011): 189–203.

Anon. "Autoimmunity in Thyroid Disease." *Lancet* 272 (15 November 1958): 1049–50.

Anon. "The Immunology of Thyroid Disease." *Lancet* 270 (25 May 1957): 1075–76.

Arrhenius, Svante. *Immunochemistry.* New York: Macmillan, 1907.

Arthus, N. M. "Injections répétées de serum du cheval chez le lapin." *Comptes Rendus des Séances de la Société de Biologie* 55 (1903): 817–20.

Avery, Andrea. "Rip Tide: Swimming through Life with Rheumatoid Arthritis." In *The Politics of Women's Bodies: Sexuality, Appearance, and Behavior,* edited by Rose Weitz, 255–67. New York: Oxford University Press, 2010.

Avery, Oswald T., Colin M. MacLeod, and Maclyn McCarty. "Induction of Transformation by a Desoxyribonucleic Acid Fraction Isolated from Pneumococcus Type III." *Journal of Experimental Medicine* 79 (1944): 137–58.

Avrameus, Stratis. "Natural Autoantibodies: From 'Horror Autotoxicus' to 'Gnothi Seauton.'" *Immunology Today* 12 (1991): 154–59.

———. "Natural Autoantibodies: Self-recognition and Physiological Autoimmunity." In *Natural Autoantibodies: Their Physiological Role and Regulatory Significance*, edited by Y. Schoenfeld and D. Isenberg, 1–15. Boca Raton: CRC Press, 1993.

Bach, Jean-François. "Type 1 Diabetes." In *The Autoimmune Diseases*. 4th ed., edited by Noel R. Rose and Ian R. Mackay, 483–500. Amsterdam: Elsevier, 2006.

Banchereau, J., and R. M. Steinman. "Dendritic Cells and the Control of Immunity." *Nature* 392 (1998): 245–52.

Barbellion, W. N. P. *The Journal of a Disappointed Man*. New York: George H. Doran, 1917.

Bayliss, W. M., and Ernest H. Starling. "The Mechanism of Pancreatic Secretion." *Journal of Physiology* 28 (1902): 325–52.

Benacerraf, Baruj. *From Caracas to Stockholm: A Life in Medical Science*. New York: Prometheus Books, 1998.

———. "Reminiscences." *Immunological Reviews* 84 (1985): 7–27.

———. "Role of MHC Gene Products in Immune Regulation." *Science* 212 (1981): 1229–38.

Benacerraf, Baruj, and Hugh O. McDevitt. "Histocompatibility-linked Immune Response Genes." *Science* 175 (1972): 273–79.

Beutner, Ernst H., Ernest Witebsky, Noel R. Rose, and Jospeh R. Gerbasi. "Localization of Thyroid and Spinal Cord Autoantibodies by Fluorescent Antibody Technic." *Proceedings of the Society for Experimental Biology and Medicine* 97 (1958): 712–16.

Bielschowsky, Marianne, B. J. Helyer, and J. B. Howie. "Spontaneous Haemolytic Anaemia in Mice of the NZB/BL Strain." *Proceedings of the University of Otago Medical School* 37 (1959): 9–11.

Billingham, Rupert E. "Reminiscences of a Transplanter." In *History of Transplantation: Thirty-Five Recollections*, edited by Paul I. Terasaki, 75–91. Los Angeles: UCLA Tissue Typing Laboratory, 1991.

Billingham, Rupert E., Leslie Brent, and P. B. Medawar. "Actively Acquired Tolerance of Foreign Cells." *Nature* 172 (1953): 603–6.

Billings, Frank. "Chronic Focal Infections and Their Etiologic Relations to Arthritis and Nephritis." *Archives of Internal Medicine* 9 (1912): 484–98.

Billings, Frank, G. H. Coleman, and W. G. Hibbs. "Chronic Infectious Arthritis: Statistical Report with End-Results." *Journal of the American Medical Association* 78 (1922): 1097–1105.

Bordet, Jules. "Anaphylaxis—Its Importance and Mechanism." *Journal of State Medicine* 21 (1913): 449–64.

Botazzo, Gian Franco, Alejo Florin-Christensen, and Deborah Doniach. "Islet Cell Antibodies in Diabetes Mellitus with Autoimmune Polyendocrine Deficiencies." *Lancet* ii (1974): 1280–83.

Bouchard, Charles Jacques. *Lectures on Autointoxication*. Translated by T. Oliver. Philadelphia: Davis, 1894.

Brass, Steven D, Howard L. Weiner, and David A. Hafler. "Multiple Sclerosis." In *The Autoimmune Diseases*. 4th ed., edited by Noel R. Rose and Ian R. Mackay, 615–32. Amsterdam: Elsevier, 2006.

Breinl, F., and F. Haurowitz. "Chemische Untersuchung des Prazipitätes aus Hämoglobin und Anti-Hämoglobin-Serum und Bemerkungen über die Natur der Antikorper." *Zeitschrift für Physiologische Chemie* 192 (1930): 45–57.

Brent, Leslie. "Tolerance and GVHD: An Exciting Decade." In *History of Transplantation*, edited by Paul I. Terasaki, 95–107. Los Angeles: UCLA Tissue Typing Laboratory, 1991.

Bretscher, Peter, and Melvin Cohn. "A Theory of Self-Nonself Discrimination." *Science* 169 (1970); 1042–49.

Briche, Gérard. *Furiculum Vitae: Chronique Hospitalière d'un Lupus*. Paris: n.p., 1979.

Bronte, Emily. *Wuthering Heights*. New York: Triangle, 1991.

Brophy, Brigid. *Baroque 'n' Roll and Other Essays*. London: Hamish Hamilton, 1987.

Brown, John. "One Man's Experience with Multiple Sclerosis." In *Multiple Sclerosis: Psychological and Social Aspects*, edited by Aart F. Simons, 21–29. London: William Heinemann, 1984.

Brown, Joseph. *Medical Essays on Fever, Inflammation, Rheumatism, Diseases of the Heart*. London, 1828.

Budd, William. *On the Causes of Fevers (1839)*. Edited by Dale C. Smith. Baltimore: Johns Hopkins University Press, 1984.

Burnet, F. Macfarlane. "Antibody Production as a Special Example of Protein Synthesis in Vivo." *Australian Journal of Science* 1 (1939): 172–73.

———. "Antibody Production in the Light of Recent Genetic Theory." *Australian Journal of Science* 8 (1946): 143–46.

———. "Autoimmune Disease." *Medical Journal of Australia* i (1962): 1–6.

———. "Autoimmune Disease, Experimental and Clinical." *Proceedings of the Royal Society of Medicine* 55 (1962): 619–26.

———. "Autoimmune Disease: Some General Principles." *Postgraduate Medicine* 30 (1961): 91–95.

———. "The Basis of Allergic Diseases." *Medical Journal of Australia* i (1948): 29–35.

———. *Biological Aspects of Infectious Disease*. Cambridge: Cambridge University Press, 1940.

———. *Cellular Immunology*. Cambridge: Cambridge University Press, 1969.

———. *Changing Patterns: An Atypical Autobiography*. Melbourne: Heinemann, 1968.

———. *The Clonal Selection Theory of Acquired Immunity.* Nashville: Vanderbilt University Press, 1959.

———. "A Darwinian Approach to Immunity." In *Molecular and Cellular Basis of Antibody Formation: Proceedings of a Symposium, Prague, June 1–5, 1965,* edited by J. Sterzl, 17–20. New York: Academic Press, 1965.

———. *Enzyme, Antigen and Virus: A Study of Macromolecular Pattern in Action.* Cambridge: Cambridge University Press, 1956.

———. "How Antibodies Are Made." *Scientific American* 191 (1954): 74–78.

———. "The Impact of Ideas on Immunology." *Cold Spring Harbor Symposia in Quantitative Biology* 32 (1967): 1–8.

———. *The Integrity of the Body: A Discussion of Modern Immunological Ideas.* Cambridge, MA: Harvard University Press, 1963.

———. "Life's Complexities: Misgivings about Models." *Australasian Annals of Medicine* 4 (1968): 36–67.

———. "The Mechanism of Immunity." *Scientific American* 204 (1961): 58–65.

———. "Medical Education and Research: Impressions of an American Visit," *Medical Journal of Australia* ii (1944): 557–62.

———. "Men or Molecules? A Tilt at Molecular Biology." *Lancet* i (1966): 37–39.

———. "A Modification of Jerne's Theory of Antibody Production Using the Concept of Clonal Selection." *Australian Journal of Science* 20 (1957): 67–69.

———. *The Production of Antibodies: A Review and a Theoretical Discussion.* Melbourne: Macmillan, 1941.

———. "The Seeds of Time: The Impact of Microbiology on Human Affairs Since Lister's Day." *Medical Journal of Australia* i (1952): 301–7.

———. *Self and Not-Self.* Cambridge: Cambridge University Press, 1969.

———. "The Theories of Antibody Formation." In *Immunity and Virus Infection,* edited by V. Najjar, 1–17. New York: John Wiley and Sons, 1959.

Burnet, F. Macfarlane, and Frank Fenner. "Genetics and Immunology." *Heredity* 2 (1948): 289–324.

———. *The Production of Antibodies,* 2nd ed. Melbourne: Macmillan, 1949.

Burnet, F. Macfarlane, and Margaret C. Holmes. "Genetic Investigations of Autoimmune Disease in Mice." *Nature* 207 (1965): 368.

Campbell, Timothy. "Interview: Roberto Esposito." Translated by Anna Papacone. *Diacritics* 36 (2006): 49–56.

Carswell, Robert. *Pathological Anatomy: Illustrations of the Elementary Forms of Disease.* London, 1838.

Cazenave, P. L. A. "Lupus érythèmateux (érythème centrifuge)," *Annales des Maladies de la Peau et de la Syphilis* 3 (1850/51): 297–99.

Chang, Timothy S., Bruce Glick, and A. R. Winter. "The Significance of the Bursa of Fabricius of Chickens in Antibody Production." *Poultry Science* 34 (1955): 1187.

Charcot, Jean-Martin. *Clinical Lectures on Senile and Chronic Diseases.* Translated by William Tulse. London, 1881.

———. "Histologie de la sclèrose en plaques." *Gazette des Hôpitaux* 41 (1868): 554–55, 557–58, 566.

Chase, Merrill W. "The Allergic State." In *Bacterial and Mycotic Infections of Man,* edited by René J. Dubos, 110–53. Philadelphia: J. B. Lippincott, 1948.

———. "The Cellular Transfer of Cutaneous Sensitivity to Tuberculin." *Proceedings of the Society of Experimental Biology and Medicine* 59 (1945): 134–35.

———. "Irreverent Recollections, from Cooke and Coca, 1928–1978." *Journal of Allergy and Clinical Immunology* 64 (1979): 306–20.

Chatenoud, Lucienne, Benoît Salomon, and Jeffrey A. Bluestone. "Suppressor T Cells—They're Back and Critical for the Regulation of Autoimmunity!" *Immunological Reviews* 182 (2001): 149–83.

Chellingsworth, Miriam C. "Multiple Sclerosis." In *When Doctors Get Sick,* edited by Harvey Mandell and Howard Spiro, 89–94. New York: Plenum Medical Book Company, 1987.

Chevassut, Kathleen. "The Aetiology of Disseminated Sclerosis." *Lancet* i (1930): 552.

Claman, Henry N., E. A. Chaperon, and R. F. Triplett. "Thymus-marrow Cell Combinations: Synergism in Antibody Production." *Proceedings of the Society of Experimental Biology and Medicine* 122 (1966): 1167–71.

Clendinnen, Inga. *Tiger's Eye: A Memoir.* New York: Scribner, 2000.

Clutterbuck, Henry. *An Enquiry into the Seat and Nature of Fever.* London, 1807.

———. *An Essay on Pyrexia and Symptomatic Fever.* London, 1837.

———. "Remarks on Dr. Elliotson's Clinical Lecture on Fever." *Lancet* 13 (21 November 1829): 274–76.

Cohen, Irun R. "Biomarkers, Self-antigens and the Immunological Homunculus." *Journal of Autoimmunity* 29 (2007): 246–49.

———. "The Cognitive Principle Challenges Clonal Selection." *Immunology Today* 13 (1992): 441–44.

———. "Kadishman's Tree, Escher's Angels, and the Immunological Homunculus." In *Autoimmunity: Physiology and Disease,* edited by Antonio Coutinho and Michael Kazatchkine, 7–18. Dordrecht: Kluwer, 1994.

———. *Tending Adam's Garden: Evolving the Cognitive Immune Self.* London: Academic Press, 2000.

Cohn, Melvin. "Commentary." In *Genetics of the Immune Response,* edited by E. Möller and G. Möller, 195. New York: Plenum Press, 1983.

———. "The Immune System: A Weapon of Mass Destruction Invented by Evolution to Even the Odds During the War of the DNAs." *Immunological Reviews* 185 (2002): 24–38.

———. "Immunology: What Are the Rules of the Game?" *Cellular Immunology* 5 (1972): 1–20.

———. "The Wisdom of Hindsight." *Annual Review of Immunology* 12 (1994): 1–62.

Cohn, Melvin, N. A. Mitchison, William E. Paul, Arthur M. Silverstein, David W. Talmage, and Martin Weigert. "Reflections on the Clonal-Selection Theory." *Nature Reviews: Immunology* 7 (2007): 823–30.

Coombs, Robin R. A., A. E. Mourant, and R. R. Race. "A New Test for Detection of Weak or 'Incomplete' Rh Agglutinins." *British Journal of Experimental Pathology* 6 (1945): 255–66.

Coons, Albert H., and M. H. Kaplan. "Localization of Antigen in Tissue Cells. II: Improvements in a Method for the Detection of Antigens by Means of Fluorescent Antibody." *Journal of Experimental Medicine* 91 (1950): 15–30.

Corrigan, D. J. *Lectures on the Nature and Treatment of Fever.* Dublin, 1853.

Corris, Peter. *Sweet and Sour: A Diabetic Life.* Lismore: Southern Cross University Press, 2000.

Coutinho, Antonio, L. Forni, D. Holmberg, F. Ivars, and N. Vaz. "From an Antigen-centered, Clonal Perspective of Immune Responses to an Organism-centered, Network Perspective of Autonomous Activity in a Self-referential Immune System." *Immunological Reviews* 79 (1984): 151–68.

Coutinho, Antonio, and Michael Kazatchkine. "Autoimmunity Today." In *Autoimmunity: Physiology and Disease,* edited by Antonio Coutinho and Michael Kazatchkine, 1–6. Dordrecht: Kluwer, 1994.

Crick, Francis H. "On Protein Synthesis." *Symposia of the Society for Experimental Biology* 12 (1958): 138–63.

Crick, Francis H. C., and James D. Watson. "Structure of Small Viruses." *Nature* 177 (1956): 473–75.

Crow, James F. "A Golden Anniversary: Cattle Twins and Immune Tolerance." *Genetics* 144 (1996): 855–59.

Cruveilhier, Jean. *Anatomie pathologique du corps humaine.* Paris: Ballière, 1829–42.

Dacie, J. V. *The Haemolytic Anaemias, Congenital and Acquired: Part II: The Auto-Immune Haemolytic Anaemias,* 2nd ed. London: J. and A. Churchill, 1962.

Dale, Henry H. "The Biological Significance of Anaphylaxis." *Proceedings of the Royal Society of London* 91 (1920): 126–46.

Dameshek, William, and Steven O. Schwartz. "The Presence of Hemolysins in Acute Hemolytic Anemia: Preliminary Note." *New England Journal of Medicine* 218 (1938): 75–80.

Dameshek, William, and Steven O. Schwartz, with Sonya Gross. "Hemolysins as the Cause of Clinical and Experimental Hemolytic Anemias, with Particular Reference to the Nature of Spherocytosis and Increased Fragility." *American Journal of Medical Science* 196 (1938): 769–92.

Darlington, C. D. "Heredity, Development and Infection." *Nature* 154 (1944): 164–69.

Dausset, Jean. *Clin d'Oeil à la Vie: La Grande Aventure HLA*. Paris: Éditions Odile Jacob, 1998.

———. "The HLA Adventure." In *History of HLA: Ten Recollections*, edited by Paul I. Terasaki, 1–19. Los Angeles: UCLA Tissue Typing Laboratory, 1990.

———. "Iso-leuco-anticorps," *Acta Haematologica* 20 (1958): 156–66.

———. "Leuco-agglutinins. IV: Leuco-agglutinins and Blood Transfusion." *Vox Sanguinis* 4 (1954): 190–98.

———. "The Major Histocompatibility Complex in Man—Past, Present, and Future Concepts." Nobel Lecture, 8 December 1980, 628–41.

Dausset, Jean, and A. Nenna. "Présence d'une leuco-agglutinine dans le sérum d'un cas d'agranulocytose chronique." *Comptes Rendus de la Société de Biologie* 146 (1952): 1539–41.

Dausset, Jean, and Felix Rapaport, "The HLA story." In *Immunology: The Making of a Modern Science*, edited by Richard B. Gallagher, Jean Gilder, G. J. V. Nossal, and Gaetano Salvatore, 111–20. London: Academic Press, 1995.

———. "The Role of Blood Group Antigens in Human Histocompatibility." *Annals of the New York Academy of Sciences* 129 (1966): 408–20.

Derrida, Jacques. "Autoimmunity: Real and Symbolic Suicides," translated by Pascale-Anne Brault and Michael Naas. In *Philosophy in a Time of Terror: Dialogues with Jürgen Habermas and Jacques Derrida*, edited by Giovanna Borradori, 85–136. Chicago: University of Chicago Press, 2003.

———. "'Eating well,' or the Calculation of the Subject." In *Points: Interviews 1974–1994*, edited by Elisabeth Weber and translated by Peggy Kamuf, 255–86. Stanford: Stanford University Press, 1995.

———. "Faith and Knowledge: Two Sources of 'Religion' at the Limits of Reason Alone [1996]". In *Religion,* edited by Jacques Derrida and Gianni Vattimo, 1–17. Cambridge: Polity Press, 1998.

———. *Rogues: Two Essays on Reason.* Translated by Pascale-Anne Brault and Michael Naas. Stanford: Stanford University Press, 2005.

———. *Specters of Marx: The State of the Debt, the Work of Mourning and the New International.* Translated by Peggy Kamuf. New York: Routledge, 1994.

Dickens, Charles. *Bleak House*. London, 1853.

———. *Great Expectations.* London, 1861.

———. *Little Dorrit.* London, 1857.

———. *The Personal History of David Copperfield.* London: Chapman and Hall, 1972.

Dienes, Louis. "The Immunological Significance of the Tuberculous Tissue." *Journal of Immunology* 15 (1928): 141–52.

Doherty, Peter C. *The Beginner's Guide to Winning the Nobel Prize: A Life in Science.* Carlton: Miegunyah Press, 2005.

———. "Challenged by Complexity: My Twentieth Century in Immunology." *Annual Reviews in Immunology* 25 (2007): 1–19.

———. "Endings and Beginnings." *Cellular and Molecular Life Sciences* 64 (2007): 1–2.
———. "Living in the Burnet Lineage." *Immunology and Cell Biology* 77 (1999): 167–76.
Donath, Julius, and Karl Landsteiner. "Über paroxysmale Hämoglobinurie." *Münchener Medizinische Wochenschrift* 51 (1904): 1590–93.
———. "Weitere Beobachtungen über paroxysmale Hämoglobinurie." *Zentralblatt für Bakteriologie* 45 (1907): 205–13.
Doniach, Deborah, and Ivan M. Roitt. "Autoimmunity in Hashimoto's Disease and Its Implications." *Journal of Clinical Endocrinology and Metabolism* 17 (1957): 1293–1304.
———. "Human Organ Specific Immunity: Personal Memories." *Autoimmunity* 1 (1988): 11–13.
Dostoyevsky, Fyodor. *The Brothers Karamazov.* Translated by Constance Garnett. London: Dent, 1927.
Draper, George, C. W. Dupertuis, and J. L. Caughe. "The Differentiation by Constitutional Methods between Pancreatic Diabetes and Diabetes of Pituitary Origin." *Transactions of the American Association of Physicians* 55 (1940): 146–53.
Edelman, Gerald, and Baruj Benacerraf. "On Structural and Functional Relations between Antibodies and Proteins of the Gamma-System." *Proceedings of the National Academy of Science* 48 (1962): 1035–42.
Edsall, Geoffrey. "What Is Immunology? A Somewhat Unscientific (and Inadequately Controlled) Enquiry into the Scope and Interests of the American Association of Immunologists and the *Journal of Immunology.*" *Journal of Immunology* 67 (1951): 167–72.
Edwards, James. "Clinical Memoranda." *British Medical Journal* i (1885): 279.
Ehrlich, Paul. "On Immunity with Special Reference to Cell Life [1900]." In *Collected Papers of Paul Ehrlich*, edited by R. Himmelweit, Martha Marquardt, and Henry Dale, 4 vols. London: Pergamon Press, 1957.
Ehrlich, Paul, and J. Morgenroth. "Fifth Communication on Hemolysis [1901]." In *Collected Papers of Paul Ehrlich*, edited by R. Himmelweit, Martha Marquardt, and Henry Dale, 4 vols. London: Pergamon Press 1957.
Elion, Gertrude B. "Historical Background of 6-mercaptopurine." *Toxicology and Industrial Health* 2 (1986): 1–9.
Elschnig, Albrecht von Graefes. "Studien zur sympatischen Ophthalmie." *Archiv für klinische und experimentelle Ophthalmologie* 75 (1910): 459–73; and 76 (1910): 509–46.
Esposito, Roberto. *Immunitas: The Protection and Negation of Life.* Translated by Zakiya Hanafi. Cambridge: Polity, 2011.
Evans, Harrison, and George T. Harding. "Multiple Sclerosis." *Diseases of the Nervous System* 1 (1940): 276–80.

Farber, Sidney, L. K. Diamond, R. D. Mercer, R. F. Sylvester, and J. A. Wolff. "Temporary Remissions in Acute Leukemia in Children Produced by Folic Acid Antagonist, 4-aminopterol-glutamic acid (aminopterin)." *New England Journal of Medicine* 238 (1948): 787–93.

Felstiner, Mary Lowenthal. *Out of Joint: A Private and Public Story of Arthritis.* Lincoln: University of Nebraska Press, 2005.

Fenner, Frank. *Nature, Nurture and Chance: The Lives of Frank and Charles Fenner.* Canberra: ANU Press, 2006.

Ferraro, Armando. "Allergic Brain Changes in Post-Scarlatina Encephalitis." *Journal of Neuropathology and Experimental Neurology* 3 (1944): 239–54.

———. "Pathology of Demyelinating Diseases as an Allergic Reaction of the Brain." *Archives of Neurology and Psychiatry* 52 (1944): 443–83.

Ferraro Armando, and G. A. Jervis, "Experimental Disseminated Encephalopathy in the Monkey," *Archives of Neurology and Psychiatry* 43 (1940): 195–209.

Ferraro, Armando, and J. E. Kilman. "Experimental Toxic Approach to Mental Disease: The Reaction of the Brain to Subcutaneous Injection of Enterogenous Toxic Substances—Indol and Histamin." *Psychiatric Quarterly* 6 (1932): 581–611.

Finlay, K. H. "Pathogenesis of Encephalitis Occurring with Vaccination, Variola and Measles." *Archives of Neurology and Psychiatry* 39 (1938): 1047–54.

Fisher, Nola. *Journeying On: An Australian Autobiography, 1947–1997.* Sydney: Nola Fisher, 1997.

Flaubert, Gustave. *Madame Bovary.* Translated by Francis Steegmuller. New York: Random House, 1957.

Fleck, Ludwik. *Genesis and Development of a Scientific Fact.* Edited by Thaddeus J. Trenn and Robert K. Merton. Translated by Fred Bradley and Thaddeus J. Trenn. Chicago: University of Chicago Press, 1979.

———. "Sérologie constitutionelle." *Travaux de la Societé des Sciences et des Lettres de Wroclaw*, series B (1956): 146–49.

———. "Some Specific Features of the Medical Way of Thinking." In *Cognition and Fact: Materials on Ludwik Fleck*, edited by Robert S. Cohen and Thomas Schelle, 39–46. Dordrecht: D. Reidel, 1986.

Fleck, Ludwik, and Laura Füllenbaum. "Clinical and Experimental Contribution on Etiology of Lupus Erythematosus." *Urologic and Cutaneous Review* 35 (1931): 358–61.

Fleck, Ludwik, and F. Goldschlag, "Experimentelle Beiträge zur Pemphigusfrage." *Klinische Wochenschrift* 16 (1937): 707–8.

———. "Further Experimental Studies of Pemphigus." *British Journal of Dermatology and Syphilis* 51 (1939): 70–76.

Fordyce, George. *Five Dissertations on Fever.* Boston: Bradford and Read, 1815.

Forsdyke, Donald R. "The Liquid Scintillation Counter as an Analogy for the Distinction between 'Self' and 'Notself' in Immunological Systems." *Lancet* 291 (1968): 281–83.

Franklin, E. C., H. R. Holman, H. J. Müller-Eberhard, and Henry G. Kunkel. "An Unusual Protein Component of High Molecular Weight in the Serum of Certain Patients with Rheumatoid Arthritis." *Journal of Experimental Medicine* 103 (1957): 425–38.

Franklin, E. C., H. G. Kunkel, and J. R. Ward. "Clinical Studies of Seven Patients with Rheumatoid Arthritis and Uniquely Large Amounts of Rheumatoid Factor." *Arthritis and Rheumatism* 1 (1958): 400–409.

Freud, Sigmund. *Beyond the Pleasure Principle.* Translated by C. J. M. Hubback. London: Hogarth Press, 1922.

Freund, Jules. "The Effect of Paraffin Oil and Mycobacteria on Antibody Formation and Sensitization," *American Journal of Clinical Pathology* 21 (1951): 645–56.

Freund, Jules, M. M. Lipton, and L. Raymond Morrison. "Demyelination in the Guinea Pig in Chronic Allergic Encephalomyelitis." *Archives of Pathology* 50 (1950): 536–39.

Freund, Jules, M. M. Lipton, and G. E. Thompson, "Aspermatogenesis in the Guinea Pig Induced by Testicular Tissue and Adjuvants." *Journal of Experimental Medicine* 97 (1953): 711–26.

Freund, Jules, and Katherine McDermott. "Sensitization to Horse Serum by Means of Adjuvants." *Proceedings of the Society of Experimental Biology and Medicine* 49 (1942): 548–53.

Freund Jules, E. R. Stern, and T. M. Pisani, "Isoallergic Encephalomyelitis and Radiculitis in Guinea Pigs after One Injection of Brain and Mycobacteria in Water-in-oil Emulsion." *Journal of Immunology* 57 (1947): 179–94.

Freyberg, R. H., C. H. Traeger, M. Patterson, W. Squires, C. H. Adams, and C. Stevenson. "Problems of Prolonged Cortisone Treatment of Rheumatoid Arthritis." *Journal of the American Medical Association* 147 (1951): 1538–43.

Friou, George J. "Antinuclear Antibodies: Diagnostic Significance and Methods." *Arthritis and Rheumatism* 10 (1967): 151–59.

———. "Clinical Application of Lupus Serum Nucleoprotein Reaction Using Fluorescent Antibody Technique." *Journal of Clinical Investigation* 36 (1957): 890–96.

Fudenberg, H. H., and F. C. Franklin. "Rheumatoid Factors and the Etiology of Rheumatoid Arthritis." *Annals of the New York Academy of Science* 124 (1965): 884–95.

Fudenberg, H. H., and H. G. Kunkel. "Physical Properties of the Red Cell Agglutinins in Acquired Hemolytic Anemia." *Journal of Experimental Medicine* 106 (1957): 689–702.

Funk, Casimir. "The Effect of a Diet of Polished Rice on the Nitrogen and Phosphorus of the Brain." *Journal of Physiology* 44 (1912): 50–53.

Gajdusek, D. Carleton. "An 'Autoimmune' Reaction against Human Tissue Antigens in Certain Acute and Chronic Diseases. I: Serological Investigations." *Archives of Internal Medicine* 10 (1958): 9–29.

———. "An 'Auto-Immune' Reaction against Human Tissue Antigens in Certain Chronic Diseases." *Nature* 179 (1957): 666–68.

Gallwey, M. Broke. "Unhealthy Inflammations." *Lancet* 1360 (22 September 1849): 307–8.

Garrod, Alfred B. "A Contribution to the Theory of the Nervous Origin of Rheumatoid Arthritis." *Medico-Chirurgical Transactions* 71 (1888): 89–105.

———. "The Great Practical Importance of Separating Rheumatoid Arthritis from Gout." *Lancet* 2012 (1892): 1033–37.

Garrod, Archibald E. *Inborn Factors in Disease* [1931]. Edited by C. R. Scriver and B. Childs. Oxford: Oxford University Press, 1989.

———. "The Incidence of Alkaptonuria: A Study in Chemical Individuality." *Lancet* 160 (1902): 1616–20.

———. *A Treatise on Gout and Rheumatic Gout (Rheumatoid Arthritis)*, 3rd ed. London, 1876.

Gepts, Willy. "Pathologic Anatomy of the Pancreas in Juvenile Diabetes Mellitus." *Diabetes* 14 (1965): 619–33.

Gershon, R. K., Philip Cohen, Ronald Hencin, and Stephen A. Liebhaber. "Suppressor T Cells." *Journal of Immunology* 108 (1972): 586–90.

Gershon, R. K., and K. Kondo. "Cell Interactions in the Induction of Tolerance: The Role of Thymic Lymphocytes." *Immunology* 18 (1970): 723–37.

———. "Infectious Immunological Tolerance." *Immunology* 21 (1971): 903–14.

Gibson, T., and P. B. Medawar. "The Fate of Skin Homografts in Man." *Journal of Anatomy* 77 (1943): 299–310.

Gilman, A., and F. S. Phillips. "The Biological Actions and Therapeutic Applications of ß-chlorethylanines and Sulfides." *Science* 103 (1946): 409–15.

Glick, Bruce. "The Saga of the Bursa of Fabricius." *BioScience* 33 (1983): 187, 190–91.

Glick, Bruce, Timothy S. Chang, and R. George Jaap. "The Bursa of Fabricius and Antibody Production in the Domestic Fowl." *Poultry Science* 35 (1956): 224–25.

Goethe, Johann Wolfgang von. *Campaign in France in the Year 1792*. Translated by Robert Farie. London, 1849.

Good, Robert A., Agustin P. Dalmasso, Carlos Martinez, Olga K. Archer, James C. Pierce, and Ben W. Papermaster. "The Role of the Thymus in Development of Immunologic Capacity in Rabbits and Mice." *Journal of Experimental Medicine* 116 (1962): 773–96.

Good, Robert A., Raymond D. A. Peterson, Carlos Martinez, David E. R. Sutherland, Michael J. Kellum, and Joanne Finstad. "The Thymus in Immunobiology, with Special Reference to Autoimmune Disease." *Annals of the New York Academy of Sciences* 124 (1965): 73–94.

Gorer, P. A. "The Antigenic Basis of Tumour Transplantation." *Journal of Pathology and Bacteriology* 47 (1938): 231–52.

———. "The Genetic and Antigenic Basis of Tumour Transplantation." *Journal of Pathology and Bacteriology* 44 (1937): 691–97.

Gorer, P. A., S. Lyman, and G. D. Snell. "Studies on the Genetic and Antigenic Basis of Tumour Transplantation: Linkage between a Histocompatibility Gene and 'Fused' in Mice." *Proceedings of the Royal Society of Biology* 135 (1948): 499–505.

Gowans, James L. "The Effect of the Continuous Re-Infusion of Lymph and Lymphocytes on the Output of Lymphocytes from the Thoracic Duct of Unanaesthetized Rats." *Journal of Experimental Pathology* 38 (1957): 67–78.

Gowans, James L., D. D. McGregor, Diana M. Cohen, and C. E. Ford. "Initiation of Immune Responses by Small Lymphocytes." *Nature* 196 (1962): 651–55.

Green, Ira, and Baruj Benacerraf. "Genetic Control of Immune Responsiveness to Limiting Doses of Proteins and Hapten Protein Conjugates in Guinea Pigs." *Journal of Immunology* 107 (1971): 374–81.

Grumet, F. Carl, Ann Conkel, Julia G. Bodmer, Walter F. Bodmer, and Hugh O. McDevitt. "Histocompatibility (HLA) Antigens Associated with Systemic Lupus Erythematosus: A Possible Genetic Predisposition to Disease." *New England Journal of Medicine* 285 (1971): 193–96.

Gundersen, Edvard. "Is Diabetes of Infectious Origin?" *Journal of Infectious Diseases* 41 (1927): 197–202.

Hammond, William Alexander. *Treatise on Diseases of the Nervous System*. New York, 1871.

Hargraves, Malcolm M. "Discovery of the L.E. Cell and Its Morphology." *Proceedings of the Staff Meetings of the Mayo Clinic* 44 (1969): 579–90.

Hargraves, Malcolm M., Helen Richmond, and Robert Morton. "Presentation of Two Bone Marrow Elements: The 'Tart' Cell and the 'L.E.' Cell." *Proceedings of the Staff Meetings of the Mayo Clinic* 23 (1948): 25–28.

Harrington, William J., Virginia Minnich, James W. Hollingsworth, and Carl V. Moore. "Demonstration of a Thrombocytopenic Factor in the Blood of Patients with Thrombocytopenic Purpura." *Journal of Laboratory and Clinical Medicine* 38 (1951): 1–10.

Haserick, John R., Lena A. Lewis, and Donald W. Bortz. "Blood Factor in Acute Disseminated Lupus Erythematosus. I: Determination of Gamma Globulin as Specific Plasma Fraction." *American Journal of Medical Sciences* 219 (1950): 660–63.

Haurowitz, Felix. "The Role of the Antigen in Antibody Formation." In *Immunity and Virus Infection,* edited by Victor A. Najjar, 18–25. New York: John Wiley and Sons, 1959.

Hegel, G. W. F. *Philosophy of Nature.* Translated by A.V. Miller. Oxford: Clarendon Press, 1970.

Heller, Joseph, and Speed Vogel. *No Laughing Matter.* New York: G. P. Putnam's Sons, 1986.

Helyer, B. J., and J. B. Howie. "Renal Disease Associated with Positive Lupus Erythematosus Tests in a Crossbred Strain of Mice." *Nature* 197 (1963): 197.

Hench, Philip S. Nobel Lecture, Nobelprize.org. Accessed 8 March 2012. http://www.nobelprize.org/nobel_prizes/medicine/laureates/1950/hench-lecture.html.

———. "Potential Reversibility of Rheumatoid Arthritis." *Proceedings of the Staff Meetings of the Mayo Clinic* 24 (1949): 167–78.

Hershey, A. D., and Martha Chase. "Independent Functions of Viral Protein and Nucleic Acid in Growth of Bacteriophage." *Journal of General Physiology* 36 (1952): 39–56.

Himsworth, H. P. "Diabetes Mellitus: Its Differentiation into Insulin-Sensitive and Insulin-Insensitive Types." *Lancet* i (1936): 127–30.

Hirszfeld, Ludwik. "Ueber die Konstitutionsserologie im Zusammenhang mit der Blutgruppenforschung." *Ergebnisse der Hygiene, Bakteriologie, Immunitätsforschung und experimentellen Therapie* 8 (1928): 367–512.

Hitchings, G. H., G. B. Elion, E. A. Falco, P. B. Russell, and H. Vander Werff. "Studies on Analogs of Purines and Pyrimidines." *Annals of the New York Academy of Science* 52 (1950): 1318–55.

Holborow, E. J., G. L. Asherson, G. D. Johnson, R. D. Barnes, and D. S. Carmichael. "Antinuclear Factor and Other Antibodies in Blood and Liver Diseases." *British Medical Journal* i (1963): 656–58.

Holman, Halsted R. "The Crisis Technology Poses for Political Institutions." *Pediatrics* 41 (1968): 322–27.

Holman, Halsted R. and Henry G. Kunkel. "Affinity between the Lupus Erythematosus Serum Factor and Cell Nuclei and Nucleoprotein." *Science* 126 (1957): 162–63.

Holmes, Margaret C., and F. Macfarlane Burnet. "The Natural History of Autoimmune Disease in NZB Mice: A Comparison with the Pattern of Human Autoimmune Manifestations." *Annals of Internal Medicine* 59 (1963): 265–76.

Howard, Mary. *Coming to Terms with Rheumatoid Arthritis.* London: Faber and Faber, 1956.

Hunter, William. *Oral Sepsis.* London: Cassell, 1901.

Hurst, Weston E. "A Review of Some Recent Observations on Demyelination." *Brain* 67 (1944): 103–24.

Hutchinson, J. "Harveian Lectures on Lupus." *British Medical Journal* i (1888): 6–10, 58–63, 113–18.

Huxley, Julian. "The Biological Basis of Individuality." *Philosophy* 1 (1926): 305–19.

———. *Essays of a Biologist.* New York: Alfred A. Knopf, 1923.

Jacobson, Denise L., Stephen J. Gange, Noel R. Rose, and Neil M. H. Graham. "Epidemiology and Estimated Population Burden of Selected Autoimmune Diseases in the United States." *Clinical Immunology and Immunopathology* 84 (1997): 233–334.

Janeway, Charles A., Jr. "Approaching the Asymptote? Evolution and Revolution in Immunology." *Cold Spring Harbor Symposium on Quantitative Biology* 54 (1989): 1–13.

———. "The Immune System Evolved to Discriminate Infectious Non-self from Non-infectious self." *Immunology Today* 13 (1992): 11–16.

———. "A Trip Through My Life with an Immunological Theme." *Annual Review of Immunology* 20 (2002): 1–28.

Jerne, Niels K. "Burnet and the Clonal Selection Theory." In *Walter and Eliza Hall Institute Annual Review: A Tribute to Sir Macfarlane Burnet,* vol. 38, 34–38. Melbourne: Exchange Press, 1979.

———. "The Complete Solution of Immunology." *Australasian Annals of Medicine* 4 (1969): 345–48.

———. "Idiotypic Networks and Other Preconceived Ideas." *Immunological Reviews* 79 (1984): 5–24.

———. "The Natural-Selection Theory of Antibody Formation." *Proceedings of the National Academy of Sciences* 41 (1955): 849–57.

———. "The Natural Selection Theory of Antibody Formation: Ten Years Later." In *Phage and the Origins of Molecular Biology,* edited by John Cairns, Gunther S. Stent, and James D. Watson, 301–12. New York: Cold Spring Harbor Laboratory Press, 1992.

———. "The Somatic Generation of Immune Recognition." *European Journal of Immunology* 1 (1971): 1–9.

———. "Summary: Waiting for the End." *Cold Spring Harbor Symposia in Quantitative Biology* 32 (1967): 591–603.

———. "Toward a Network Theory of the Immune System." *Annales d'Immunologie* 125C (1974): 373–89.

Jones, C. Handfield. "General Considerations Respecting Fever." *British Medical Journal* ii (1858): 644–45.

———. "Pathological and Therapeutical Considerations Relative to Inflammation and Fever." *British Medical Journal* ii (1859): 955–58.

Joske, R. A., and W. E. King. "The 'L.E. Cell' Phenomenon in Active Chronic Viral Hepatitis." *Lancet* ii (1955): 477–80.

Kabat, Elvin A. "Getting Started 50 Years Ago—Experiences, Perspectives, and Problems of the First 21 years." *Annual Review of Immunology* 1 (1983): 1–32.

Kabat, Elvin A., Abner Wolf, and Ada E. Bezer. "Rapid Production of Acute Disseminated Encephalomyelitis in Rhesus Monkeys by Injection of Brain Tissue with Adjuvants." *Science* 104 (1946): 362–63.

———. "The Rapid Production of Acute Disseminated Encephalomyelitis in Rhesus Monkeys by Injection of Heterologous and Homologous Brain Tissue with Adjuvants." *Journal of Experimental Medicine* 85 (1947): 117–29.

———. "Studies on Acute Disseminated Encephalomyelitis Produced Experimentally in Rhesus Monkeys. IV: Disseminated Encephalomyelitis Produced in Monkeys with Their Own Brain Tissue." *Journal of Experimental Medicine* 89 (1949): 395–98.

———. "Studies on Acute Disseminated Encephalomyelitis Produced Experimentally in Rhesus Monkeys. VII: The Effect of Cortisone." *Journal of Immunology* 68 (1952): 265–75.

Kang, Insoo, and Joe Craft. "Systemic Lupus Erythematosus: Immunologic Features." In *The Autoimmune Diseases,* 4th ed., edited by Noel R. Rose and Ian R. Mackay, 357–67. Amsterdam: Elsevier, 2006.

Kaposi, Moriz. "Neue Beiträge zur Kenntniss des Lupus Erythematosus." *Archiv für Dermatologie und Syphilologie* 4 (1872): 36–78.

Kendall, Edward C. *Cortisone: Memoirs of a Hormone Hunter.* New York: Scribner's, 1971.

Kennedy, F. "Allergy and Its Effects on the Central Nervous System." *Journal of Nervous and Mental Diseases* 88 (1938): 91–98.

Kierkegaard, Søren. *Sickness Unto Death.* Edited and translated by Howard B. Hong and Edna H. Hong. Princeton: Princeton University Press, 1980.

Kinley, Anne E. "MS: From Shock to Acceptance." *American Journal of Nursing* 80 (1980): 274–75.

Klemperer, Paul. "The Concept of Collagen Diseases." *American Journal of Pathology* 26 (1950): 505–19.

———. "The Pathogenesis of Lupus Erythematosus and Allied Conditions." *Annals of Internal Medicine* 28 (1948): 1–11.

Klemperer, Paul, Abou D. Pollack, and George Baehr. "Diffuse Collagen Disease." *Journal of the American Medical Association* 119 (1942): 331–32.

Kunkel, Henry G. "Accomplishments." *Advances in Experimental Medicine and Biology* 216A (1987): v–viii.

Kunkel, Henry G., E. H. Ahrens, W. J. Eisenmenger, A. M. Bongiovanni, and R. J. Slater. "Extreme Hypergammaglobulinemia in Young Women with Liver Disease of Unknown Etiology." *Journal of Clinical Investigation* 30 (1951): 654.

Lacy, P. E., and P. H. Wright. "Allergic Interstitial Pancreatitis in Rats Injected with Guinea Pig Serum." *Diabetes* 14 (1965): 634–42.

Lafferty, K. J., and A. Cunningham. "A New Analysis of Allogenic Interactions." *Australian Journal of Experimental Biology and Medical Science* 53 (1975): 27–42.

Lancereaux, E. "Le diabète maigre: ses symptoms, son evolution, son prognostie et son traitement." *Un Med. Paris* 20 (1880): 205–11.

Landsteiner, Karl. *Die Spezifizität der serologischen Reaktionen.* Berlin: Springer, 1933.

———. *The Specificity of Serological Reactions.* New York: Dover Publications, 1962.

Landsteiner, Karl, and Merrill W. Chase. "Experiments on Transfer of Cutaneous Sensitivity to Simple Compounds." *Proceedings of the Society of Experimental Biology and Medicine* 49 (1942): 688–90.

———. "Studies in the Sensitization of Animals with Simple Chemical Compounds. VI: Experiments on the Sensitization of Guinea Pigs to Poison Ivy." *Journal of Experimental Medicine* 69 (1939): 767–84.

Landsteiner, Karl, C. Levaditi, and E. Prásek. "Étude expérimentale du pemphigus infectieux aigu." *Comptes Rendus de la Société de Biologie*, Paris 70 (1911): 643–45.

Landsteiner, Karl, R. Müller, and O. Poetzl. "Zur Frage der Komplementbindungsreaktionen bei Syphilis." *Wiener Klinische Wockenschrift* 20 (1907): 1565–67.

Langman, R. E., and M. Cohn. "A Minimal Model for the Self-Nonself Discrimination: A Return to the Basics." *Seminars in Immunology* 12 (2000): 189–95.

Lederberg, Joshua. "Genes and Antibodies." *Science* 129 (1959): 1649–53.

———. "Ontogeny of the Clonal Selection Theory of Antibody Formation: Reflections on Darwin and Ehrlich." *Annals of the New York Academy of Sciences* 546 (1988): 175–82.

Levine, B. B., A. Ojeda, and Baruj Benacerraf. "Studies on Artificial Antigens. III: The Genetic Control of the Immune Response to Hapten-poly-L-lysine Conjugates in Guinea Pigs." *Journal of Experimental Medicine* 118 (1963): 953.

Lewis, Sinclair. *Arrowsmith*. New York: New American Library, 1961.

Liebig, Justus. *Animal Chemistry, or Chemistry in Its Applications to Physiology and Pathology*. Edited by W. Gregory. Cambridge, 1842.

Lister, Joseph, "The Relations of Clinical Medicine to Modern Scientific Development." *British Medical Journal* ii (1896): 33–41, 733, 734.

Locke, John. *An Essay Concerning Human Understanding*. Edited by Peter H. Nidditch. Oxford: Clarendon Press, 1979.

Loeb, Jacques. *The Mechanistic Conception of Life*. Cambridge, MA: Harvard University Press, 1964.

Loeb, Leo. "The Biological Basis of Individuality." *Science* 86 (1937): 1–5.

———. *The Biological Basis of Individuality*. Springfield, IL: Charles C. Thomas, 1945.

Lowry, Florence. "One Woman's Experience with Multiple Sclerosis." In *Multiple Sclerosis: Psychological and Social Aspects*, edited by Aart F. Simons, 30–35. London: William Heinemann, 1984.

Lwoff, André. "The Concept of the Virus." *Journal of General Microbiology* 17 (1957): 239–53.

Maccalister, Donald, "Gulstonian Lectures on the Nature of Fever." *Lancet* i (12 March 1887): 507–11.

Mackay, Ian R. "The Etiopathogenesis of Autoimmunity." *Seminars in Liver Disease* 25 (2005): 239–50.

Mackay, Ian R., and F. Macfarlane Burnet. *Autoimmune Diseases: Pathogenesis, Chemistry and Therapy*. Springfield, IL: Charles C. Thomas, 1963.

Mackay, Ian R., and D. Carleton Gajdusek. "An 'Autoimmune' Reaction against Human Tissue Antigens in Certain Acute and Chronic Diseases. II: Clinical Correlations." *Archives of Internal Medicine* 10 (1958): 30–46.

Mackay, Ian R., Lois Larkin, and F. Macfarlane Burnet, "Failure of 'Autoimmune' Antibody to React with Antigen Prepared from the Individual's Own Tissues." *Lancet* 270 (20 July 1957): 122–23.

Mackay, Ian R., and P. J. Morris. "Association of Autoimmune Active Chronic Hepatitis with HL-A1, 8." *Lancet* ii (1972): 793–95.

Mackay, Ian R., L. I. Taft, and D. C. Cowling. "Lupoid Hepatitis." *Lancet* ii (1956): 1323–26.

Mackay, Ian R., S. Weiden, and J. Hasker. "Autoimmune Hepatitis." *Annals of the New York Academy of Sciences* 124 (1965): 767–80.

Mackay, Ian R., S. Wieden, and B. Ungar. "Treatment of Chronic Active Hepatitis and Lupoid Hepatitis with 6-Mercaptopurine and Azathioprine." *Lancet* i (1964): 899–902.

Mackenzie, Stephen. "Paroxysmal Haemoglobinuria, with Remarks on Its Nature." *Lancet* ii (1879): 116–17, 155–57.

Mackie, Sarah, Mark Quinn, and Paul Emery. "Rheumatoid Arthritis." In *The Autoimmune Diseases*, 4th ed., edited by Noel R. Rose and Ian R. Mackay, 417–36. Amsterdam: Elsevier, 2006.

Maclean, L. D., S. J. Zak, R. L. Varco, and R. A. Good. "Thymic Tumor and Acquired Agammaglobulinemia—a Clinical and Experimental Study of the Immune Response." *Surgery* 40 (1956): 1010–17.

Maddocks, Ian. "The Effects of Prolonged Corticosteroid Therapy: An Appraisal of 66 Cases." *Australasian Annals of Medicine* 10 (1961): 223–29.

Manguso, Sarah. *The Two Kinds of Decay: A Memoir.* New York: Farrar, Straus and Giroux, 2008.

Marie, P. "Sclèrose en plaques et maladies infectieuses." *Progrès Médical Paris* 12 (1884): 287–89, 305–7, 349–51, 365–66.

Martin, Emily. "The End of the Body?" *American Ethnologist* 16 (1989): 121–40.

———. *Flexible Bodies: Tracking Immunity in American Culture—From the Days of Polio to the Age of AIDS.* Boston: Beacon Press, 1994.

———. "Toward an Anthropology of Immunology: The Body as a Nation State." *Medical Anthropology Quarterly* 4 (1990): 410–26.

Martineau, Harriet. *Life in the Sickroom: Essays by an Invalid.* London, 1844.

Masugi, Matazo. "Zur Pathogenese der diffusen Glomeronephritis als allergischer Erkrankung der Niere." *Klinische Wochenschrift* 14 (1935).

Matzinger, Polly. "The Danger Model: A Renewed Sense of Self." *Science* 296 (2002): 301–5.

———. "The Danger Model in Its Historical Context." *Scandinavian Journal of Immunology* 54 (2001): 4–9.

———. "The Evolution of the Danger Theory." *Expert Reviews in Clinical Immunology* 8 (2012): 311–17.

———. "Tolerance, Danger, and the Extended Family." *Annual Review of Immunology* 12 (1994): 991–1045.

Mauss, Marcel. "A Category of the Human Mind: The Notion of Person; The Notion of Self [1937]." Translated by W. D. Halls. In *The Category of the Person: Anthropology, Philosophy, History,* edited by Michael Carrithers, Steven Collins, and Steven Lukes, 1–25. Cambridge: Cambridge University Press, 1985.

McDevitt, Hugh O. "Discovering the Role of the Major Histocompatibility Complex in the Immune Response." *Annual Review of Immunology* 18 (2000): 1–17.

McDevitt, Hugh O., and Walter F. Bodmer. "Histocompatibility Antigens, Immune Responsiveness and Susceptibility to Disease." *American Journal of Medicine* 52 (1972): 1–8.

———. "HL-A, Immune-response Genes and Disease." *Lancet* i (1974): 1269–75.

McDevitt, Hugh O., and A. Chinitz. "Genetic Control of the Antibody Response: Relationship between Immune Response and Histocompatibility (H-2) Type." *Science* 163 (1969): 1207–8.

Medawar, P. B. "The Behavior and Fate of Skin Autografts and Skin Homografts in Rabbits." *Journal of Anatomy* 78 (1944): 176–99.

———. "Burnet and Immunological Tolerance." In *Walter and Eliza Hall Institute of Medical Research Annual Review: A Tribute to Sir Macfarlane Burnet,* 31–33. Melbourne: Exchange Press, 1979.

———. *Induction and Intuition in Scientific Thought.* Philadelphia: American Philosophical Society, 1969.

———. *Memoir of a Thinking Radish: An Autobiography.* Oxford: Oxford University Press, 1986.

———. "A Second Study of the Behavior and Fate of Skin Homografts in Rabbits." *Journal of Anatomy* 79 (1945): 157–76.

Metalnikoff, Serge. "Études sur la spermotoxine." *Annales de l'Institut Pasteur, Paris* 14 (1900): 577–89.

Metchnikoff, Élie. "Concerning the Relationship between Phagocytes and Anthrax Bacilli." *Journal of Infectious Diseases* 6 (1884): 761–70.

———. *Lectures on the Comparative Pathology of Inflammation.* London, 1893.

———. *The Prolongation of Life.* Translated by P. C. Mitchell. New York: Putnam, 1908.

———. "A Yeast of Daphnia: A Contribution to the Theory of the Struggle of Phagocytes Against Pathogens [1884]." In *Three Centuries of Microbiology,* edited by Hubert A. Lechevalier and Morris Solotorovsky, 188–95. New York: Dover, 1974.

Miescher, Peter, and M. Fauconnet. "L'absorption du facteur 'L.E.' par des noyaux cellulaires isolés." *Experimentia* 10 (1954): 252–54.

Miller, H., D. J. Newell, and A. Ridley. "Multiple Sclerosis: Treatment of Acute Exacerbations with Corticotropin (ACTH)." *Lancet* ii (1961): 1120–22.

Miller, J. F. A. P. "Antigen-specific Clone Selection *In Vivo*." *Immunology and Cell Biology* 86 (2008): 24–25.

———. "Discovering the Origins of Immunological Competence." *Annual Review of Immunology* 17 (1999): 1–17.

———. "The Discovery of the Immunological Function of the Thymus." *Immunology Today* 12 (1991): 42–45.

———. "The Discovery of Thymus Function and of the Thymus-Derived Lymphocytes." *Immunological Reviews* 185 (2002): 7–14.

———. "Effect of Neonatal Thymectomy on the Immunological Responsiveness of the Mouse." *Proceedings of the Royal Society of London* 156B (1962): 410–28.

———. "Immunological Function of the Thymus." *Lancet* ii (1961): 748–49.

Miller, J. F. A. P., and Graham F. Mitchell. "Cell to Cell Interaction in the Immune Response. I: Hemolysin-forming Cells in Neonatally Thymectomized Mice Reconstituted with Thymus or Thoracic Duct Lymphocytes." *Journal of Experimental Medicine* 128 (1968): 801–20.

Miller, J. F. A. P., and John Sprent. "Cell to Cell Interaction in the Immune Response. VI: Contribution of Thymus-Derived and Antibody-Forming Cell Precursors to Immunological Memory." *Journal of Experimental Medicine* 134 (1971): 66–82.

Mitchell, Graham F. "Selection, Memory and Selective Memories: T Cells, B Cells and Sir Mac 1968." *Immunology and Cell Biology* 86 (2008): 26–30.

Mitchell, Graham F., and J. F. A. P. Miller. "Cell to Cell Interaction in the Immune Response. II: The Source of Hemolysin-forming Cells in Irradiated Mice Given Bone Marrow and Thymus or Thoracic Duct Lymphocytes." *Journal of Experimental Medicine* 128 (1968): 821–37.

———. "Immunological Activity of the Thymus and Thoracic-duct Lymphocytes." *Proceedings of the National Academy of Sciences* 59 (1968): 296–303.

Möller, Goran. "Do Suppressor T Cells Exist?" *Scandinavian Journal of Immunology* 27 (1988): 247–50.

Moore, W. Withers. "On the Production of Heat in Fever." *British Medical Journal,* i (1884): 258–60.

Moreschi, Carlo. "Neue Tatsachen über die Blutkörperchenagglutination." *Zentralblatt für Bakteriologie* 46 (1908): 49–51.

Morgan, Isabel M. "Allergic Encephalomyelitis in Monkeys in Response to Injection of Normal Monkey Nervous Tissue." *Journal of Experimental Medicine* 85 (1947): 131–40.

Morrison, L. Raymond. "Disseminated Encephalomyelitis Experimentally Produced by the Use of Homologous Antigen." *Archives of Neurology and Psychiatry* 58 (1947): 391–416.

Morse, Erik. "Something in the Air: Interview with Peter Sloterdijk." *Frieze* 129 (2009). Accessed 29 March 2013, www.frieze.com/issue/something_in_the_air.

Morse, John Lovett. "Diabetes in Infancy and Childhood." *Boston Medical and Surgical Journal* 168 (1913): 530–35.

Moxon, Walter. "Case of Insular Sclerosis of Brain and Spinal Cord." *Lancet* i (1873): 236.

———. "Considerations Bearing on Our Present Knowledge of Fever." *Lancet* 3092 (1882): 931–33.

Mudd, S. "A Hypothetical Mechanism of Antibody Formation." *Journal of Immunology* 23 (1932): 423–27.

Murphy, J. B. *The Lymphocyte in Resistance to Tissue Grafting, Malignant Disease, and Tuberculous Infection: An Experimental Study*, Rockefeller Institute of Medical Research Monograph No. 21. New York: Rockefeller Institute, 1926.

Naito, S., N. Namerow, M. R. Mickey, and P. I. Terasaki. "Multiple Sclerosis: Association with HL-A3." *Tissue Antigens* 2 (1972): 1–4.

Nancy, Jean-Luc. *Corpus*. Translated by Richard A. Rand. New York: Fordham University Press, 2008.

Napier, A. David. *The Age of Immunology: Conceiving a Future in an Alienating World*. Chicago: University of Chicago Press, 2003.

———. "Non Self Help: How Immunology Might Reframe the Enlightenment." *Cultural Anthropology* 27 (2012): 122–37.

Nepom, G. T., and H. Erlich. "MHC Class II Molecules and Autoimmunity." *Annual Review of Immunology* 9 (1991): 493–525.

Nerup, J., P. Platz, O. Ortved Andersen, M. Christy, *et al.* "HL-A Antigens and Diabetes Mellitus." *Lancet* ii (1974): 864–66.

Nicholls, E. H. and F. L. Richardson. "Arthritis Deformans." *Journal of Medical Research* 21 (1909): 149–222.

Nietzsche, Friedrich. "On the Uses and Disadvantages of History for Life." In *Untimely Meditations*, translated by R. J. Hollingdale. Cambridge: Cambridge University Press, 1997.

Nossal, G. J. V. "Burnet and Science—An Appreciation." *Australasian Annals of Medicine* 4 (1969): 311–15.

———. "The Coming of Age of the Clonal Selection Theory." In *Immunology 1930–1980*, edited by P. M. H. Mazumdar, 41–48. Toronto: Wall and Thompson, 1989.

———. "One Cell—One Antibody: Impact on Immunological Theory and Practice." In *The Immunological Revolution: Facts and Witnesses*, edited by Andor Szentivanyi and Herman Friedman, 81–89. Boca Raton: CRC Press, 1994.

Nossal, G. J. V., and Joshua Lederberg. "Antibody Production by Single Cells." *Nature* 181 (1958): 1419–20.

O'Connor, Flannery. *The Habit of Being: Letters*. Edited by Sally Fitzgerald. New York: Farrar, Strauss and Giroux, 1979.

Osler, William. "On the Visceral Complications of Erythema Exudativum Multiforme." *American Journal of Medical Science* 110 (1895): 629–46.

———. "On the Visceral Manifestations of the Erythema Group of Skin Diseases." *Transactions of the Association of American Physicians* 18 (1903): 599–624.

Owen, Ray D. "Immunogenetic Consequences of Vascular Anastomoses between Bovine Twins." *Science* 102 (1943): 400–401.

Oxford Centre for Diabetes, Endocrinology and Metabolism (OCDEM). "Diabetes Stories, Personal Tales of Diabetes Through the Decades," Interviews Nos. 8, 31, 7. Accessed 20 August 2013. http://www.diabetes-stories.com/index.asp.

Papermaster, B. W., D. I. Friedman, and R. A. Good. "Relationships of the Bursa of Fabricius to Immunological Responsiveness and Homograft Immunity in Chickens." *Proceedings of the Society for Experimental Biology and Medicine* 110 (1962): 62–64.

Pauling, Linus. "A Theory of the Structure and Process of Formation of Antibodies." *Journal of the American Chemical Society* 62 (1940): 2643–57.

Payne, Rose. "Early History of HLA." In *History of HLA: Ten Recollections*, edited by Paul I. Terasaki, 21–32. Los Angeles: UCLA Tissue Typing Laboratory, 1990.

Pirquet, Clemens von. "Allergie," *Münchener Medizinische Wochenschrift* 30 (1906): 1457–58.

———. *Allergy.* Chicago: American Medical Association, 1911.

Pirquet, Clemens von, and Béla Schick. *Serum Sickness.* Translated by Béla Schick. Baltimore: Williams and Wilkins, 1951.

Poe, Edgar Allan. "For Annie." In *The Collected Works of Edgar Allan Poe,* 3 vols., edited by Thomas Ollive Mabbott. Cambridge, MA: Belknap Press, 1969.

Polley, H. F., and C. H. Slocumb. "Behind the Scenes with Cortisone and ACTH." *Mayo Clinic Proceedings* 51 (1976): 471–77.

Poynton, F. J. and A. Paine. "The Etiology of Rheumatic Fever." *Lancet* ii (1900): 861–69, 932–35.

Putnam, Tracy J. "The Pathogenesis of Multiple Sclerosis: A Possible Vascular Factor." *New England Journal of Medicine* 209 (1933): 786–90.

———. "Studies in Multiple Sclerosis. IV: 'Encephalitis' and Sclerotic Plaques Produced by Venular Obstruction." *Archives of Neurology and Psychiatry* 33 (1935): 929–40.

———. "Studies in Multiple Sclerosis. VIII: Etiological Factors in Multiple Sclerosis." *Annals of Internal Medicine* 9 (1936): 854–63.

———. "Venous Thrombosis as the Primary Alteration in the Lesions of 'Encephalomyelitis' and Multiple Sclerosis." *New England Journal of Medicine* 216 (1937): 103–4.

Putnam, Tracy J., J. B. McKenna, and L. R. Morrison. "Studies in Multiple Sclerosis. I: The Histogenesis of Experimental Sclerotic Plaques and Their Rela-

tion to Multiple Sclerosis," *Journal of the American Medical Association* 97 (1931): 1591–95.
Quastler, Henry. "The Measure of Specificity." In *Essays on the Use of Information Theory in Biology*, edited by Henry Quastler, 41–71. Urbana: University of Illinois Press, 1953.
Rachford, B. K. "A Case of Diabetes Mellitus in a Child Five Years Old." *New York Medical Journal* 2 (1889): 629.
Renold, A. E., J. S. Soeldner, and J. Steinke. "Immunological Studies with Homologous and Heterologous Pancreatic Insulin in the Cow." In *Aetiology of Diabetes and Its Complications*, edited by Margaret P. Cameron and Maeve O'Connor, 122–39. Ciba Foundation Colloquia, vol. 15. London: Churchill, 1964.
Ribatti, D., E. Crivellato, and A. Vacca. "The Contribution of Bruce Glick to the Definition of the Role Played by the Bursa of Fabricius in the Development of the B Cell Lineage." *Clinical and Experimental Immunology* 145 (2006): 1–4.
Richet, Charles. "Anaphylaxis." *Scandinavian Journal of Immunology* 31 (1990): 375–88.
———. "Ancient Humoralism and Modern Humoralism." *British Medical Journal* ii (1910): 921–26.
———. *L'Anaphylaxie.* Paris: Librairie Felix Alcan, 1912.
Richet, Charles, and Paul Portier. "De l'action anaphylactique de certains venins." *Comptes Rendus des Séances de la Société de Biologie* 54 (1902): 170–72.
Rivers, T. M. "Viruses." *Journal of the American Medical Association* 92 (1929): 1147–52.
Rivers, T. M., and F. F. Schwentker. "Encephalomyelitis Accompanied by Myelin Destruction Experimentally Produced in Monkeys." *Journal of Experimental Medicine* 61 (1935): 689–702.
Rivers, T. M., D. H. Sprunt, and G. P. Berry. "Observations on Attempts to Produce Acute Disseminated Encephalomyelitis in Monkeys." *Journal of Experimental Medicine* 58 (1933): 39–53.
Robbins, W. C. "Recollections of Henry G. Kunkel and the Rockefeller Institute for Medical Research, 1954–1957." *Lupus* 12 (2003): 218–21.
Robbins, W. C., H. R. Holman, Helmut Deicher, and Henry G. Kunkel. "Complement Fixation with Cell Nuclei and DNA in Lupus Erythematosus." *Proceedings of the Society for Experimental Biology and Medicine* 96 (1957): 575–79.
Roitt, I. M., Deborah Doniach, P. N. Campbell, and R. Vaughan Hudson. "Auto-Antibodies in Hashimoto's Disease (Lympadenoid Goitre)." *Lancet* ii (1956): 820–21.
Roitt, I. M., M. F. Greaves, G. Torrigiani, J. Bostoff, and J. H. L. Playfair. "The Cellular Basis of Immunological Responses." *Lancet* ii (1969): 367–71.
Rood, J. J. van, and A. van Leeuwen. "The HLA Story as Seen from Leiden." In *History of HLA*, edited by Paul I. Terasaki, 33–59. Los Angeles: UCLA Tissue Typing Laboratory, 1990.

Rose, Harry M., Charles Ragan, Elizabeth Pearce, and Miriam Olmstead Lipman. "Differential Agglutination of Normal and Sensitized Sheep Erythrocytes by Sera of Patients with Rheumatoid Arthritis." *Proceedings of the Society for Experimental Biology and Medicine* 68 (1948): 1–6.

Rose, Noel R. "Autoimmunity: A Personal Memoir" *Autoimmunity* 1 (1988): 15–21.

———. "The Discovery of Thyroid Autoimmunity." *Immunology Today.* 12 (1991): 167–68.

———. "Life Among the Contrivances." *Nature Immunology* 7 (2006): 1009–11.

———. "Reflections on Tolerance, Self-Tolerance and Felix Milgrom." *Transplantation Proceedings* 31 (1999): 1460–63.

———. "Thyroid Autoimmunity: A Voyage of Discovery." In *The Immunologic Revolution: Facts and Witnesses,* edited by Andor Szentivanyi and Herman Friedman. Boca Raton: CRC Press, 1994.

Rose, Noel R., and Constantin Bona. "Defining Criteria for Autoimmune Diseases (Witebsky's Postulates Revisited)." *Immunology Today* 14 (1993): 426–30.

Rose, Noel R., and Ian R. Mackay. "Molecular Mimicry: A Critical Look at Exemplary Instances in Human Diseases." *Cellular and Molecular Life Sciences* 57 (2000): 542–51.

Rose, Noel R., and Ernest Witebsky. "Studies on Organ Specificity. V: Changes in Thyroid Glands of Rabbits Following Active Immunization with Rabbit Thyroid Extracts." *Journal of Immunology* 76 (1956): 417–27.

Rubinstein, Renate. *Take It and Leave It: Aspects of Being Ill.* Translated by Karin Fierke and Aad Janssen. London: Marion Boyars, 1989.

Saint, E. G., W. E. King, R. A. Joske, and E. S. Finckh. "The Course of Infectious Hepatitis with Special Reference to Prognosis and the Chronic Stage." *Australasian Annals of Medicine* 2 (1953): 113–27.

Sakaguchi, Shimon. "Naturally Arising CD4 Regulatory T Cells for Immunologic Self Tolerance and Negative Control of Immune Responses." *Annual Review of Immunology* 22 (2004): 531–62.

Salmon, D. E., and T. Smith. "On a New Method of Inducing Immunity to Contagious Diseases." *Proceedings of the Biological Society of Washington* 3 (1884–86): 29–33.

Schick, Béla. "Pediatrics in Vienna at the Turn of the Century." *Journal of Pediatrics* 50 (1957): 114–24.

Schwartz, R., and W. Dameshek. "Drug-induced Immunological Tolerance." *Nature* 183 (1959): 1682–83.

Schwentker, Francis F., and Frank C. Comploir. "The Production of Kidney Antibodies by Injection of Homologous Kidney Plus Bacterial Toxins." *Journal of Experimental Medicine* 70 (1939): 223–30.

Schwentker, Francis F., and T. M. Rivers. "The Antibody Response of Rabbits to Injections of Emulsions of Homologous Brain." *Journal of Experimental Medicine* 60 (1934): 559–74.

Scriver, C. R., and B. Childs, eds. *Inborn Factors in Disease* [1931]. Oxford: Oxford University Press, 1989.

Searle, Henry. "On the Nature of Inflammation and Irritation." *Lancet* 25 (26 September 1835): 26–29.

Sercarz, E., A. Oki, and G. Gammon. "Central Versus Peripheral Tolerance: Clonal Inactivation Versus Suppressor T Cells, the Second Half of the 'Thirty Years War.'" *Immunology* Supplement 2 (1989): 9–14.

Sever, John L. "Major Technological Advances Affecting Clinical and Diagnostic Immunology." *Clinical and Diagnostic Laboratory Immunology* 4 (1997): 1–3.

Seyle, Hans. "Allergy and the General Adaptation Syndrome." *International Archives of Allergy* 3 (1952): 267–78.

Shaw, George Bernard. "Preface on the Doctors." In *The Doctor's Dilemma*. Harmondsworth: Penguin, 1979.

Silverstein, Arthur M., and Noel R. Rose. "On the Mystique of the Immunological Self." *Immunological Reviews* 159 (1997): 197–206.

———. "There Is Only One Immune System! The View from Immunopathology." *Seminars in Immunology* 12 (2000): 173–78.

Simonson, Morten. "Graft-versus-host Reactions: The History that Never Was, and the Way Things Happened to Happen." *Immunological Reviews* 88 (1985): 5–23.

———. "The Impact on the Developing Embryo and Newborn Animal of Adult Homologous Cells." *Acta Pathologica et Microbiologica Scandinavica* 40 (1957): 480–500.

Sloterdijk, Peter. *Sphären III. Plurale sphärologie. Shäume.* Frankfurt-am-Main: Suhrkamp Verlag, 2004.

Sloterdijk, Peter, and H.-J. Heinrichs. *Die Sonne und der Tod.* Frankfurt-am-Main: Suhrkamp Verlag, 2006.

Smadel, Joseph E. "Experimental Nephritis in Rats Induced by Injections of Antikidney Serum Preparation and Immunological Studies of Nephrotoxin." *Journal of Experimental Medicine* 64 (1936): 921–42.

Smadel, Joseph E., and H. F. Swift. "Reverse Anaphylaxis in Rats, with Special Attention to Kidney Damage." *Journal of Immunology* 32 (1937): 75–81.

Snell, George D. "A Geneticist's Recollections of Early Transplantation Studies." In *A History of Transplantation: Thirty-Five Recollections*, edited by Paul I. Terasaki, 19–36. Los Angeles: UCLA Tissue Typing Laboratory, 1991.

———. ed. *Biology of the Laboratory Mouse.* New York: Dover, 1941.

Soulillou, Jean-Paul. "An Interview with Jean Dausset." *American Journal of Transplantation* 4 (2004): 4–7.

Starling, Ernest Henry. "The Wisdom of the Body." *British Medical Journal* ii (1923): 685–90.

Steinman, R. M., and Z. A. Cohn. "Identification of a Novel Cell Type in Peripheral Lymphoid Organs of Mice. I: Morphology, Quantification, Tissue Distribution." *Journal of Experimental Medicine* 137 (1973): 1142–62.

Stephens, Mrs Leslie. *Notes From Sick Rooms.* London: 1883.

Stevenson, Lewis D., and Ellsworth C. Alvord, "Allergy in the Nervous System," *American Journal of Medicine* 3 (1947): 614–20.

Sullivan, J. W. N. "The New Scientific Horizon." *Nation and Athenaeum* 29 (13 August 1921): 722.

Swift, Homer. "The Nature of Rheumatic Fever." *Journal of Laboratory and Clinical Medicine* 21 (1936): 551–63.

———. "Rheumatic Fever." *Journal of the American Medical Association* 92 (1929): 2071–83.

Talmage, David W. "Allergy and Immunology." *Annual Review of Medicine* 8 (1957): 239–56.

———. "Origin of the Cell Selection Theories of Antibody Formation." In *Immunology: The Making of a Modern Science*, edited by Richard B. Gallagher, Jean Gilder, G. J. V. Nossal, and Gaetano Salvatore, 23–38. London: Academic Press, 1995.

Tan, E. M., and Henry G. Kunkel. "Characteristics of a Soluble Nuclear Antigen Precipitating with Sera of Patients with Systemic Lupus Erythematosus." *Journal of Immunology* 96 (1966): 464–71.

Taylor, Mark C. "The Betrayal of the Body: Live Not." In *Nots*, 214–55. Chicago: University of Chicago Press, 1993.

Thomson, John. *Lectures on Inflammation.* Edinburgh, 1813.

Traub, Erich. "Persistence of Lymphocytic Chorio-Meningitis Virus in Immune Animals and Its Relation to Immunity." *Journal of Experimental Medicine* 63 (1936): 847–61.

Travers, Benjamin. *The Physiology of Inflammation and the Healing Process.* London, 1844.

Turner, W. J. *The Duchess of Popocatapetl.* London: J. M. Dent and Sons, 1939.

Tweedie, Alexander. *Clinical Illustrations of Fever.* London, 1830.

———. "Lectures on Fevers. X: Mortality of Continued Fever." *Lancet* 1916 (19 May 1860): 485–89.

Valéry, Paul. *Cahiers*, 2 vols., edited by Judith Robinson. Vol. 2 [1944–45]. Paris: Gallimard, 1974.

Vallery-Radot, Pasteur, and V. Heimann. *Hypersensibilités specifiques dans les affections cutanées: Anaphylaxie, idiosyncrasie.* Paris: Masson et Cie, 1930.

Vance, Russell E. "A Copernican Revolution? Doubts About the Danger Theory." *Journal of Immunology* 165 (2000): 1725–28.

Varela, Francisco J. "Organism: A Meshwork of Selfless Selves." In *Organism and the Origins of Self*, edited by Alfred I. Tauber, 79–107. Dordrecht: Kluwer, 1991.

Varela, Francisco J., and Mark R. Anspach. "The Body Thinks: The Immune System and the Process of Somatic Individuation." In *Materialities of Communication,* edited by Hans Ulrich Gumbrecht and K. Ludwig Pfeiffer, translated by William Whobrey, 273–85. Stanford: Stanford University Press, 1994.

Vaz, N. M., and Francisco J. Varela. "Self and Nonsense: An Organism-centered Approach to Immunology." *Medical Hypotheses* 4 (1978): 231–67.

Vesell, Elliot S. "Recollections from the Kunkel Laboratory, 1956–1958." *Lupus* 12 (2008): 238–41.

Virchow, Rudolf. "Cellular Pathology [1855]." In *Disease, Life, and Man: Selected Essays by Rudolf Virchow*. Edited and translated by Lelland J. Rather. Stanford: Stanford University Press, 1958.

———. "One Hundred Years of General Pathology [1895]." In *Disease, Life and Man: Selected Essays by Rudolf Virchow*. Edited and translated by Lelland J. Rather. Stanford: Stanford University Press, 1958.

———. "Recent Progress in Science and Its Influence on Medicine and Surgery [1898]." In *Disease, Life and Man: Selected Essays by Rudolf Virchow*. Edited and translated by Lelland J. Rather. Stanford: Stanford University Press, 1957.

———. "Standpoints in Scientific Medicine [1847]." In *Disease, Life, and Man: Selected Essays by Rudolf Virchow*, Edited and translated by Lelland J. Rather. Stanford: Stanford University Press, 1958.

Vladutiu, Adrian O., and Noel R. Rose. "Autoimmune Murine Thyroiditis Relation to Histocompatibility (H-2) Type." *Science* 174 (1971): 1137–39.

Waaler, Erik. "On the Occurrence of a Factor in Human Serum Activating the Specific Agglutination of Sheep Blood Corpuscles." *Acta Pathologica Microbiologica Scandinavia* 17 (1940): 172–88.

Waksman, Byron H. "The Etiology of Rheumatic Fever: A Review of Theories and Evidence." *Medicine: Analytic Reviews of General Medicine and Applied Immunology* 28 (1949): 143–200.

———. "Experimental Allergic Encephalomyelitis and the 'Auto-allergic' Diseases." *International Archives of Allergy and Applied Immunology* 14 (Supplement) (1959): 1–87.

———. "Further Studies of Skin Reactions in Rabbits with Experimental Allergic Encephalomyelitis." *Journal of Infectious Diseases* 99 (1956): 258–69.

Waksman, Byron H., and L. Raymond Morrison, "Tuberculin Type Sensitivity to Spinal Cord Antigen in Rabbits with Allergic Encephalomyelitis," *Journal of Immunology* 66 (1951): 421–44.

Waksman, Byron H., and Joost J. Oppenheim. "The Contribution of the Cytokine Concept to Immunology." In *Immunology: The Making of a Modern Science*, edited by Richard B. Gallagher, Jean Gilder, G. J. V. Nossal, and Gaetano Salvatore, 133–34. London: Academic Press, 1995.

Waldenström, Jan G. "Leber, Blutproteine und Nahrungseiweiss." *Deutsche Geschichte Verdau Stoffwechselkrankeit* 15 (1950): 113–19.

Warner N. L., and A. Szenberg. "Dissociation of Immunological Responsiveness in Fowls with a Hormonally Arrested Development of Lymphoid Tissue." *Nature* 194 (1962): 146–47.

———. "Effect of Neonatal Thymectomy on the Immune Response of the Chicken." *Nature* 196 (1962): 784–85.

Warner, N. L., A. Szenberg, and F. Macfarlane Burnet. "The Immunological Role of Different Lymphoid Organs in the Chicken. I: Dissociation of Immunological Responsiveness." *Australasian Journal of Experimental Biology and Medical Science* 40 (1962): 373–88.

Wassermann, August von, Albert Neisser, C. Bruck, and A. Schucht. "Weitere Mitteilungen über den Nachweis spezifischluetischer Substanzen durch Komplementverankerung." *Zeitschrift für Hygiene und Infektionskrankheiten* 55 (1906): 451–77.

Watson, James D., and Francis H. C. Crick. "Molecular Structure of Nucleic Acids: A Structure for Deoxyribose Nucleic Acid," *Nature* 171 (1953): 737–38.

Weigle, William O. "Immunological Unresponsiveness." *Advances in Immunology* 16 (1973): 61–123.

Weil, Edmund. "Das Problem der Serologie der Lues in der Darstellung Wassermanns." *Berliner Klinische Wochenschrift* 58 (1921): 966–70.

Weil, Edmund, and H. Braun. "Ueber das Wesen der luetischen Erkrankung auf Grund der neueren Untersuchungen," *Wiener Klinische Wochenschrift* 22 (1909): 372–74.

Weiner, Norbert. *The Human Use of Human Beings: Cybernetics and Society.* London: Sphere Books, 1968.

Wells, H. G. "Introduction." In W. N. P. Barbellion, *The Journal of a Disappointed Man*. New York: George H. Doran, 1917.

———. *The War of the Worlds.* London: William Heinemann, 1898.

Whitehead, Alfred North. *Process and Reality: An Essay in Cosmology.* Corrected edition, edited by D. Griffin and D. Sherborne. New York: Free Press, 1978.

———. *Science and the Modern World.* New York: Free Press, 1997.

Whittingham, Senga, and Ian R. Mackay. "Design and Functions of a Department of Clinical Immunology." *Clinical and Experimental Immunology* 8 (1971): 857–61.

———. "The 'Pemphigus' Antibody and the Immunopathies Affecting the Thymus." *British Journal of Dermatology* 84 (1971): 1–6.

Whittingham, Senga, Ian R. Mackay, and Z. S. Kiss. "An Interplay of Genetic and Environmental Factors in Familial Hepatitis and Myasthenia Gravis." *Gut* 11 (1970): 811–16.

Wilson, Horace. "Diabetes in a Young Child." *Lancet* ii (1886): 1376.

Winkelman, N. W., and Matthew T. Moore. "Allergy and Nervous Diseases." *Journal of Nervous and Mental Diseases* 93 (1941): 736–49.

Witebsky, Ernest. "Historical Roots of Present Concepts of Immunopathology." In *Immunopathology, First International Symposium, Basel/Seelisberg, 1958,* edited by Pierre Grubar and Peter Miescher, 1–13. Basel: Benno Schwabe, 1958.

Witebsky, Ernest, and Noel R. Rose. "Studies on Organ Specificity. IV: Production of Rabbit Thyroid Antibodies in the Rabbit." *Journal of Immunology* 76 (1956): 408–16.

Witebsky, Ernest, Noel R. Rose, Kornel Terplan, John R. Paine, and Richard W. Egan. "Chronic Thyroiditis and Autoimmunization." *Journal of the American Medical Association* 164 (1957): 1439–47.

Wolstenholme, G. E. W., and J. Knight, eds. *The Immunologically Competent Cell: Its Nature and Origin.* Ciba Foundation Study Group No. 16. London: Churchill, 1963.

Wood, Ian J. *Discovery and Healing in Peace and War: An Autobiography.* Port Melbourne: Riall Print, 1984.

Wood, Ian J., and Leon I. Taft. *Diffuse Lesions of the Stomach: An Account with Special Reference to the Value of Gastric Biopsy.* London: Edward Arnold, 1958.

Woolf, Virginia. "8 December 1921." In *The Diary of Virginia Woolf.* Edited by A. O. Bell and A. McNeillie, vol. 2. New York: Harcourt Brace, 1980.

Wunderlich, Carl August. *On the Temperature in Disease: A Manual of Medical Thermometry.* London 1871.

Yidirim-Toruner, Cagri, and Betty Diamond. "Current and Novel Therapeutics in Treatment of SLE." *Journal of Allergy and Clinical Immunology* 127 (2011): 303–14.

Young, Lawrence E., G. Miller, and R. M. Christian. "Clinical and Laboratory Observations on Autoimmune Hemolytic Disease." *Annals of Internal Medicine* 35 (1951): 507–17.

Zandman-Goddard, G., and Y. Schoenfeld. "HIV and Autoimmunity." *Autoimmunity Reviews* 1 (2002): 329–37.

Ziegler, John L., and Daniel P. Stites. "Hypothesis: AIDS Is an Autoimmune Disease Directed at the Immune System and Triggered by a Lymphotropic Retrovirus." *Clinical Immunology and Immunopathology* 41 (1986): 305–13.

Zimmerman, H. J., P. Heller, and R. P. Hill, "Extreme Hyperglobulinemia in Subacute Hepatic Necrosis." *New England Journal of Medicine* 244 (1951): 245–49.

Zinkernagel, Rolf M. "About the Discovery of MHC-Restricted T Cell Recognition." In *Immunology: The Making of a Modern Science*, edited by Richard B. Gallagher, Jean Gilder, G. J. V. Nossal, and Gaetano Salvatore, 85–94. London: Academic Press, 1995.

———. "Uncertainties—Discrepancies in Immunology." *Immunological Reviews* 185 (2002): 103–25.

Zinkernagel, Rolf M., and Peter C. Doherty. "Immunological Surveillance Against Altered Self Components by Sensitized T Lymphocytes in Lymphocytic Choriomeningitis." *Nature* 251 (1974): 547–48.

———. "MHC-restricted Cytotoxic T Cells: Studies on the Biologic Role of Polymorphic Major Transplantation Antigens Determining T Cell Restriction,

Specificity, Function and Responsiveness." *Advances in Immunology* 27 (1979): 52–177.

———. "Restriction of *in vitro* T Cell-mediated Cytotoxicity in Lymphocytic Chorio-meningitis within a Syngeneic or Semiallogeneic System," *Nature* 248 (1974): 701–2.

Zinsser, Hans. *As I Remember Him: The Biography of R.S.* Boston: Little, Brown, 1940.

———. "On the Significance of Bacterial Allergy in Infectious Diseases." *Bulletin of the New York Academy of Medicine* 4 (1928): 351–83.

———. *Rats, Lice and History.* Boston: Little, Brown, 1935.

Zinsser, Hans, Hugh K. Ward, and Frederick B. Jennings. "The Significance of Bacterial Allergy as a Sign of Resistance." *Journal of Immunology* 10 (1925): 719–23.

Secondary Sources

Ackerknecht, Erwin H. "Broussais, or a Forgotten Medical Revolution." *Bulletin of the History of Medicine* 27 (1953): 320–43.

———. "Diathesis: The Word and the Concept in Medical History." *Bulletin of the History of Medicine* 56 (1982): 317–25.

———. *Medicine at the Paris Hospital, 1794–1848.* Baltimore: Johns Hopkins University Press, 1967.

Ada, Gordon L., and Gustav Nossal. "The Clonal-Selection Theory." *Scientific American* 257 (1987): 62–69.

Agamben, Giorgio. *Homo Sacer: Sovereign Power and Bare Life.* Translated by Daniel Heller-Roazen. Stanford: Stanford University Press, 1998.

Anderson, Warwick. *The Collectors of Lost Souls: Turning Kuru Scientists into Whitemen.* Baltimore: Johns Hopkins University Press, 2008.

———. "The Military Spur to Australian Medical Research." *Health and History* 15 (2013): 80–103.

———. "Natural Histories of Infectious Disease: Ecological Vision in Twentieth-Century Biomedical Science." *Osiris* 19 (2004): 39–61.

———. "The Reasoning of the Strongest: The Polemics of Skill and Science in Diagnosis." *Social Studies of Science* 22 (1992): 653–84.

Anderson, Warwick, and Cecily Hunter. " 'Wars Have an Overflow on Everything': Interview with Ian R. Mackay and Patricia Mackay." *Health and History* 15 (2013): 104–17.

Anderson, Warwick, and Ian R. Mackay. "Fashioning the Immunological Self: The Biological Individuality of F. Macfarlane Burnet." *Journal of the History of Biology* 47 (2014): 147–75.

———. "Gut Reactions—From Celiac Affection to Autoimmune Model." *New England Journal of Medicine* 371 (2014): 6–7.

Anderson, Warwick, Barbara Gutmann Rosenkrantz, and Miles Jackson. "Toward an Unnatural History of Immunology." *Journal of the History of Biology* 27 (1994): 575–94.

Apple, Rima D. *Vitamania: Vitamins in American Culture.* New Brunswick: Rutgers University Press, 1996.

Bailin, Miriam. *The Sickroom in Victorian Fiction: The Art of Being Ill.* Cambridge: Cambridge University Press, 1994.

Baxter, Alan G. *Germ Warfare: Breakthroughs in Immunology.* St. Leonards: Allen and Unwin, 2000.

Beller, Steven. *Vienna and the Jews, 1867–1938: A Cultural History.* Cambridge: Cambridge University Press, 1991.

Belt, Henk van den, and Bart Gremmen. "Specificity in the Era of Koch and Ehrlich: A Generalized Interpretation of Ludwik Fleck's 'Serological' Thought Style." *Studies in History and Philosophy of Science* 21 (1990): 463–79.

Benacerraf, Baruj. "Henry G. Kunkel, 1916–1983: An Appreciation of the Man and His Scientific Contributions." *Journal of Experimental Medicine* 161 (1985): 878–80.

Benedek, Thomas G. "Historical Background of Discoid and Systemic Lupus Erythematosus." In *Dubois' Lupus Erythematosus,* edited by Daniel Jeffrey Wallace, Bevra Hahn, and Edmund L. Dubois, 2–15. New York: Williams and Wilkins, 2007.

———. "The History of Bacteriologic Concepts of Rheumatic Fever and Rheumatoid Arthritis." *Seminars in Arthritis and Rheumatology* 36 (2006): 109–23.

Benison, Saul. *Tom Rivers: Reflections on a Life in Medicine and Science.* Cambridge, MA: MIT Press, 1967.

Benton, Graham. " 'And Dying Thus Around Us Every Day': Pathology, Ontology and the Discourse of the Diseased Body, a Study of Illness and Contagion in *Bleak House.*" *Dickens Quarterly* 11 (1994): 69–80.

Bernard, Jean, Marcel Bessis, and Claude Debru, eds. *Soi et Non-Soi.* Paris: Éditions du Seuil, 1990.

Bialynicki-Birula, Rafal. "The 100th Anniversary of the Wasserman-Neisser-Bruck Reaction." *Clinics in Dermatology* 26 (2008): 79–88.

Bliss, Michael. *The Discovery of Insulin.* Chicago: University of Chicago Press, 2007.

Blustein, Bonnie E. *Preserve Your Love for Science: Life of William Alexander Hammond, American Neurologist.* New York: Cambridge University Press, 1991.

Boonen, Annelies, Jan van de Rest, Jan Dequeker, and Sjef van der Linden. "How Renoir Coped with Rheumatoid Arthritis." *British Medical Journal* 315 (1997): 1704–8.

Bradshaw, David. "The Best of Companions: J. W. N. Sullivan, Aldous Huxley, and the New Physics." *Review of English Studies* 47 (1996): 188–206.

Braidotti, Rosi. *Nomadic Subjects: Embodiment and Sexual Difference in Contemporary Feminist Theory.* New York: Columbia University Press, 1994.

Brent, Leslie. *A History of Transplantation Immunology.* San Diego: Academic Press, 1997.

———. "Sir Peter Brian Medawar." *Proceedings of the American Philosophical Society* 136 (1992): 439–41.

Brooks, Nancy A., and Ronald R. Matson. "Social-Psychological Adjustment to Multiple Sclerosis: A Longitudinal Study." *Social Science and Medicine* 16 (1982): 2129–35.

Buklijas, Tatjana. "Surgery and National Identity in Late Nineteenth-Century Vienna." *Studies in the History and Philosophy of Biological and Biomedical Sciences* 38 (2007): 756–74.

Bulloch, William. *History of Bacteriology.* London: Oxford University Press, 1938.

Burgio, G. R. "Biological Individuality and Disease: From Garrod's *Chemical Individuality* to HLA-associated Diseases." *Acta Biotheoretica* 41 (1993): 219–30.

Burkitt, Ian. "The Shifting Concept of the Self." *History of the Human Sciences* 7, 1994: 7–28.

Burnet, F. Macfarlane. *Walter and Eliza Hall Institute, 1915–1965.* Melbourne: Melbourne University Press, 1971.

Bury, Michael. "Chronic Illness as Biographical Disruption." *Sociology of Health and Illness* 4 (1982): 167–82.

Bynum, W. F. "Cullen and the Study of Fevers in Britain, 1760–1820." *Medical History*, Supplement 1, (1981): 135–47.

———. *Science and the Practice of Medicine in the Nineteenth Century.* Cambridge: Cambridge University Press, 1994.

Bywaters, E. G. L. "Historical Aspects of the Aetiology of Rheumatoid Arthritis." *British Journal of Rheumatology* 27 (1988) Suppl. 2: 110–15.

Campbell, Timothy. "'Bios,' Immunity, Life: The Thought of Roberto Esposito." *Diacritics* 36 (2006): 2–22.

Canguilhem, Georges. *The Normal and the Pathological.* Translated by Carolyn R. Fawcett with Robert S. Cohen. New York: Zone Books, 1989.

Cantor, David. "Cortisone and the Politics of Drama, 1949–55." In *Medical Innovations in Historical Perspective,* edited by John V. Pickstone, 165–84. Basingstoke, Hampshire: Macmillan, 1992.

Carroy, Jacqueline. "Playing with Signatures: The Young Charles Richet." In *The Mind of Modernism: Medicine, Psychology, and the Cultural Arts in Europe and America, 1880–1940,* edited by Mark Micale, 217–49. Stanford: Stanford University Press, 2004.

Caspar, Stephen T. "Trust, Protocol, Gender, and Power in Interwar British Biomedical Research: Kathleen Chevassut and the 'Germ' of Multiple Sclerosis." *Journal of the History of Medicine and Allied Sciences* 66 (2010): 180–215.

Chadarevian, Soraya de, and Harmke Kamminga, eds. *Molecularizing Biology and Medicine: New Practices and Alliances, 1910s–1970s.* Amsterdam: Harwood Academic, 1998.

Charlesworth, Max, Lyndsay Farrell, Terry Stokes, and David Turnbull. *Life Among the Scientists: An Anthropological Study of an Australian Scientific Community.* Melbourne: Oxford University Press, 1989.

Charmaz, Kathy. "Loss of Self: A Fundamental Form of Suffering in the Chronically Ill." *Sociology of Health and Illness* 5 (1983): 168–95.

———. "Struggling for a Self: Identity Levels of the Chronically Ill." In *Experience and Management of Chronic Illness,* edited by Julius A. Roth and Peter Conrad, 283–321. Greenwich, Conn., JAI Press.

Chernyak, Leon, and Alfred I. Tauber. "The Idea of Immunity: Metchnikoff's Metaphysics and Science." *Journal of the History of Biology* 23 (1990): 187–249.

Christensen, Allan Conrad. *Nineteenth-Century Narratives of Contagion: "Our Feverish Contract."* London: Routledge, 2005.

Coen, Deborah R. "Living Precisely in Fin-de-Siècle Vienna." *Journal of the History of Biology* 39 (2006): 493–523.

Cohen, Ed. *A Body Worth Defending: Immunity, Biopolitics, and the Apotheosis of the Modern Body.* Durham, NC: Duke University Press, 2009.

Cohen-Cole, Jamie. "The Creative American: Cold War Salons, Social Science, and the Cure for Modern Society." *Isis* 100 (2010): 219–62.

Corbin, Juliet, and Anselm L. Strauss. "Accompaniments of Chronic Illness: Change in Body, Self, Biography, and Biographical Time." In *The Experience and Management of Chronic Illness,* edited by Julius A. Roth and Peter Conrad, 249–81. Greenwich, CT: Jai Press, 1987.

Corner, George W. *A History of the Rockefeller Institute, 1901–1953: Origins and Growth.* New York: Rockefeller Institute Press, 1964.

Creager, Angela N. H. "Adaptation or Selection? Old Issues and New Stakes in the Postwar Debates over Bacterial Drug Resistance." *Studies in History and Philosophy of Biological and Biomedical Sciences* 38 (2007): 159–90.

Crist, Eileen, and Alfred I. Tauber. "Debating Humoral Immunity and Epistemology: The Rivalry of Immunochemists Jules Bordet and Paul Ehrlich." *Journal of the History of Biology* 30 (1997): 321–56.

———. "Selfhood, Immunity, and the Biological Imagination: The Thought of Frank Macfarlane Burnet." *Journal of the History of Biology* 15 (1999): 509–33.

Cruse, Julius M., and Robert E. Lewis, Jr. "David W. Talmage and the Advent of the Cell Selection Theory of Antibody Synthesis." *Journal of Immunology* 152 (1994): 919–29.

Davis, Vivianne de Vahl. "A History of the Walter and Eliza Hall Institute of Medical Research, 1915–1978: An Examination of the Personalities, Politics, Finances, Social Relations and Scientific Organization of the Hall Institute." Ph.D. thesis, University of New South Wales, 1979.

Douglas, C., and Margaret Cyr. "The History of Lupus Erythematosus from Hippocrates to Osler." *Rheumatic Disease Clinics of North America* 14 (1988): 1–14.

Dubos, René. *Louis Pasteur: Free Lance of Science.* London: Gollancz, 1951.

Elzen, Boelie. "Two Ultracentrifuges: A Comparative Study of the Social Construction of Artifacts." *Social Studies of Science* 16 (1986): 621–62.

Engelhardt, Dietrich von. "Hegel's Philosophical Understanding of Illness." In *Hegel and the Sciences*, Boston Studies in the Philosophy of Science, no. 64, edited by Robert S. Cohen and Marx W. Wartofsky, 41–54. Dordrecht: D. Reidel, 1984.

Fenner, F. J. "Frank Macfarlane Burnet, 3 September 1899–31 August 1985." *Biographical Memoirs of Fellows of the Royal Society* 33 (1987): 101–62.

Feudtner, Christopher. *Bittersweet: Diabetes, Insulin and the Transformation of Illness.* Durham: University of North Carolina Press, 2003.

Firth, Douglas. "The Case of Augustus d'Este (1794–1848): The First Account of Disseminated Sclerosis." *Proceedings of the Royal Society of Medicine* 34 (1941): 381–84.

Forsdyke, Donald R. "Immunology (1955–1975): The Natural Selection Theory, the Two Signal Hypothesis and Positive Repertoire Selection." *Journal of the History of Biology* 45 (2012): 139–61.

Foster, W. D. *A History of Medical Bacteriology and Immunology.* London: W. Heinemann, 1970.

Foucault, Michel. *The Birth of the Clinic: An Archeology of Medical Perception.* Translated by A. M. Sheridan. London: Tavistock, 1973.

———. *Discipline and Punish: The Birth of the Prison.* Translated by Alan Sheridan. New York: Pantheon, 1978.

———. "Introduction." In Georges Canguilhem, *The Normal and the Pathological*, 7–24. Translated by Carolyn R. Fawcett with Robert S. Cohen. New York: Zone Books, 1989.

———. *"Society Must be Defended" Lectures at the Collège de France, 1975–1976.* Translated by David Macey. New York: Picador, 2003.

Fox, Renée. *Experiment Perilous: Physicians and Patients Facing the Unknown.* New Brunswick: Transaction Publishers, 1998.

Fraser, Kevin J. "The Waaler-Rose Test: Anatomy of the Eponym," *Seminars in Arthritis and Rheumatism* 18 (1988): 61–71.

Frederickson, Sten, and Slavenka Kam-Hansen. "The 150-year Anniversary of Multiple Sclerosis: Does Its Early History Give an Etiological Clue?" *Perspectives in Biology and Medicine* 32 (1989): 237–43.

Gale, Edwin A. M. "The Discovery of Type 1 Diabetes." *Diabetes* 50 (2001): 217–26.

———. "The Rise of Childhood Type I Diabetes in the Twentieth Century." *Diabetes* 51 (2002): 3353–61.

Gaudillière, Jean-Paul. *Inventer la Biomédicine: La France, l'Amerique et la Production des Savoirs du Vivant (1945–1965).* Paris: Éditions la Découverte, 2002.

———. "Mapping as Technology: Genes, Mutant Mice, and Biomedical Research." In *Classical Genetic Research and Its Legacy: The Mapping Cultures of Twentieth-Century Genetics*, edited by Jean-Paul Gaudillière and Hans-Jörg Rheinberger, 173–204. New York: Routledge, 2004.

Gerr, Werner, and Christophe Viret. "From Variation in Genetic Information to Clonal Deletion: Joshua Lederberg's Immunological Legacy." *Immunology and Cell Biology* 87 (2009): 264–66.
Giddens, Anthony. *Modernity and Self-Identity: Self and Society in the Late Modern Age.* Cambridge: Polity, 1991.
Gieson, Gerald L. *The Private Science of Louis Pasteur.* Princeton: Princeton University Press, 1995.
Glyn, John. "The Discovery and Early Use of Cortisone." *Journal of the Royal Society of Medicine* 91 (1998): 513–17.
Goffman, Erving. *Asylums.* New York: Doubleday, 1971.
Gould, Stephen Jay. "The Hardening of the Modern Synthesis." In *Dimensions of Darwinism: Themes and Counterthemes in Twentieth-Century Evolutionary Theory,* edited by Marjorie Grene, 71–93. Cambridge: Cambridge University Press, 1983.
Gowans, James L. "Peter Medawar: His Life and Work." *Immunology Letters* 21 (1989): 5–8.
Gradmann, Christoph. *Laboratory Disease: Robert Koch's Medical Bacteriology.* Translated by Elborg Forster. Baltimore: Johns Hopkins University Press, 2009.
Grosz, Elizabeth. *Volatile Bodies: Toward a Corporeal Feminism.* St. Leonards: Allen and Unwin, 1994.
Gurney, Michael S. "Disease as Device: The Role of Smallpox in Bleak House." *Literature and Medicine* 9 (1990): 79–92.
Gurr, Werner, and Christophe Viret. "From Variation in Genetic Information to Clonal Deletion: Joshua Lederberg's Immunological Legacy." *Immunology and Cell Biology* 87 (2009): 264–66.
Gussow, Zachary, and George S. Tracy. "The Role of Self-Help Clubs in Adaptation to Chronic Illness and Disability." *Social Science and Medicine* 10 (1976): 407–14.
Hacohen, Malachi Haim. "The Culture of Viennese Science and the Riddle of Austrian Liberalism." *Modern Intellectual History* 6 (2009): 369–96.
Haraway, Donna. "The Biopolitics of Postmodern Bodies: Constitutions of Self in Immune System Discourse." In *Simians, Cyborgs, and Women: The Reinvention of Nature.* New York: Routledge, 1991.
———. "The High Cost of Information in Post–World War II Evolutionary Biology: Ergonomics, Semiotics, and the Sociobiology of Communication Systems." *Philosophical Forum* 13 (1981–82): 244–78.
Harington, Anne. *Reenchanted Science: Holism in German Culture from Wilhelm II to Hitler.* Princeton: Princeton University Press, 1999.
Harvey, A. McGehee. "Clinical Science at the Mayo Clinic: The Concept of Team Research." In *Science at the Bedside: Clinical Research in American Medicine,* edited by A. McGehee Harvey, 368–92. Baltimore: Johns Hopkins University Press, 1981.

———. *Science at the Bedside: Clinical Research in American Medicine* Baltimore: Johns Hopkins University Press, 1981.

Heller, Thomas S., Morton Sosna, and David E. Wellburn, eds. *Reconstructing Individualism: Autonomy, Individuality, and the Self in Western Thought.* Stanford: Stanford University Press, 1986.

Hess, Volker. "Standardizing Body Temperature: Quantification in Hospitals and Daily Life, 1850–1900." In *Body Counts: Medical Quantification in Historical and Sociological Perspective,* edited by Gérard Jorland, A. Opinel, and George Weisz, 109–26. Montreal: McGill-Queens University Press, 2005.

Hettenyi, G., and J. Karsch. "Cortisone Therapy: A Challenge to Academic Medicine in 1949–1952." *Perspectives in Biology and Medicine* 40 (1997): 426–39.

Hickey, William F. "The Pathology of Multiple Sclerosis: A Historical Perspective." *Journal of Neuroimmunology* 98 (1999): 37–44.

Hobbins, Peter. "'Immunization Is as Popular as a Death Adder': The Bundaberg Tragedy and the Politics of Medical Science in Inter-war Australia." *Social History of Medicine* 24 (2011): 426–44.

Hooker, Claire. "Diphtheria, Immunization and the Bundaberg Tragedy: A Study of Public Health in Australia." *Health and History* 2 (2000): 52–78.

Hughes, Everett C. *The Sociological Eye.* Chicago: Aldine, 1971.

Jackson, Mark. *Allergy: The History of a Modern Malady.* London: Reaktion Books, 2006.

———. "John Freeman, Hay Fever and the Origins of Clinical Allergy in Britain, 1900–1950." *Studies in History and Philosophy of Biological and Biomedical Sciences* 34 (2003): 473–90.

Jamieson, Michelle. "Imagining 'Reactivity': Allergy Within the History of Immunology." *Studies in History and Philosophy of the Biological and Biomedical Sciences* 41 (2010): 356–66.

Janik, Allan, and Stephen Toulmin. *Wittgenstein's Vienna.* New York: Simon and Schuster, 1973.

Jellinek, E. H. "Heine's Illness: The Case for Multiple Sclerosis." *Journal of the Royal Society of Medicine* 83 (1990): 516–19.

Kay, Lily E. "Cybernetics, Information, Life: The Emergence of Scriptural Representations of Heredity." *Configurations* 5 (1997): 23–91.

———. "Laboratory Technology and Biological Knowledge: The Tiselius Electrophoresis Apparatus, 1930–1945." *History and Philosophy of the Life Sciences* 10 (1988): 51–72.

———. *The Molecular Vision of Life: Caltech, the Rockefeller Foundation, and the Rise of the New Biology.* New York: Oxford University Press, 1993.

Keating, Peter. "Georges Canguilhem's *The Normal and the Pathological*: A Restatement and a Commentary." In *Singular Selves: Historical Issues and*

Contemporary Debates in Immunology, edited by Anne Marie Moulin and Alberto Cambrosio. Amsterdam: Elsevier, 2001.

———. "Holistic Bacteriology: Ludwik Hirszfeld's Doctrine of Serogenesis between the Two World Wars." In *Greater than the Parts: Holism in Biomedicine, 1920–1950*, edited by Christopher Lawrence and George Weisz, 283–302. New York: Oxford University Press, 1998.

Keating, Peter, and Alberto Cambrosio. "Helpers and Suppressors: On Fictional Characters in Medicine." *Journal of the History of Biology* 30 (1997): 381–96.

Keating, Peter, Alberto Cambrosio, and Michael Mackenzie. "The Tools of the Discipline: Standards, Models, and Measures in the Affinity/Avidity Controversy in Immunology." In *The Right Tools for the Job: At Work in Twentieth-Century Life Sciences*, edited by Adele E. Clarke and Joan H. Fujimura, 312–54. Princeton: Princeton University Press, 1992.

Keller, Evelyn Fox. *Making Sense of Life: Explaining Biological Development with Models, Metaphors, and Machines*. Cambridge, MA: Harvard University Press, 2002.

Klein, Jan. *Immunology: The Science of Self-Nonself Discrimination*. New York: John Wiley and Sons, 1982.

———. "Seeds of Time: Fifty Years Ago Peter A. Gorer Discovered the H2 Complex." *Immunogenetics* 24 (1986): 331–38.

Kleinman, Arthur. *The Illness Narratives: Suffering, Healing, and the Human Condition*. New York: Basic Books, 1988.

Koselleck, Reinhard. *The Practice of Conceptual History: Timing History, Spacing Concepts*. Translated by Todd Presner, Kerstin Hehnke, and Jobst Welge. Stanford: Stanford University Press, 2002.

Kroker, Kenton. "Immunity and Its Other: The Anaphylactic Selves of Charles Richet." *Studies in History and Philosophy of Biological and Biomedical Sciences* 30 (1999): 273–96.

Lawrence, Christopher, and J. Dixey, "Practising on Principle: Joseph Lister and the Germ Theory of Disease." In *Medical Theory, Surgical Practice: Studies in the History of Surgery*, edited by Christopher Lawrence, 153–215. London: Routledge, 1992.

Lawrence, Christopher, and George Weisz, eds. *Greater than the Parts: Holism in Biomedicine, 1920–1950*. New York: Oxford University Press, 1998.

Lederer, Susan E. *Subjected to Science: Human Experimentation in America before the Second World War*. Baltimore: Johns Hopkins University Press, 1995.

Lesky, Erna. *The Vienna Medical School of the Nineteenth Century*. Translated by L. Williams and I. S. Levij. Baltimore: Johns Hopkins University Press, 1976.

———. "Viennese Serological Research about the Year 1900: Its Contribution to the Development of Clinical Medicine." *Bulletin of the New York Academy of Medicine* 49 (1973): 100–111.

———. "Wassermann and the Vienna School of Serology." *International Journal of Dermatology* 16 (1977): 526–30.

Liebenau, Jonathan. *Medical Science and Medical Industry: The Formation of the American Pharmaceutical Industry.* Baltimore: Johns Hopkins University Press, 1987.

López-Beltrán, Carlos. "Forging Heredity: From Metaphor to Cause, A Reification Story." *Studies in the History and Philosophy of Science* 25 (1995): 211–35.

Louie, J. S. "Renoir, His Art, and His Arthritis." In *Art, History and Antiquity of Rheumatic Diseases,* edited by T. Appleboom. Brussels: Elsevier, 1987.

Löwy, Ilana. *Between Bench and Bedside: Science, Healing, and Interleukin-2 in a Cancer Ward.* Cambridge, MA: Harvard University Press, 1997.

———. "The Epistemology of the Science of an Epistemologist of the Sciences: Ludwik Fleck's Professional Outlook and Its Relationship to His Philosophical Works." In *Cognition and Fact: Materials on Ludwik Fleck,* edited by Robert S. Cohen and Thomas Schelle, 421–42. Dordrecht: D. Reidel, 1986.

———. "The Immunological Constitution of the Self." In *Organism and the Origins of Self,* edited by Alfred I. Tauber, 43–75. Dordrecht: Kluwer, 1991.

———. "Immunology and AIDS: Growing Explanations and Developing Instruments." In *Growing Explanations: Historical Perspectives on Recent Science,* edited by M. Norton Wise, 222–47. Durham, NC: Duke University Press, 2004.

———. "Immunology and Literature in the Early Twentieth Century: *Arrowsmith* and *The Doctor's Dilemma.*" *Medical History* 32 (1988): 314–32.

———. "Immunology in the Clinics: Reductionism, Holism, or Both?" In *Crafting Immunology: Working Histories of Clinical Immunology,* edited by Kenton Kroker, Pauline M. H. Mazumdar, and Jennifer E. Keelan, 165–76. London: Ashgate, 2008.

———. "The Impact of Medical Practice on Biomedical Research: The Case of Human Leucocyte Antigens Studies." *Minerva* 25 (1987): 171–200.

———. "On Guinea Pigs, Dogs and Men: Anaphylaxis and the Study of Biological Individuality, 1902–1939." *Studies in History and Philosophy of Biological and Biomedical Sciences* 34 (2003): 399–423.

———. *The Polish School of Philosophy of Medicine: From Tytus Chalubinski (1820–1889) to Ludwik Fleck (1896–1961).* Dordrecht: Kluwer Academic, 1990.

———. "The Strength of Loose Concepts—Boundary Concepts, Federative Experimental Strategies and Disciplinary Growth: The Case of Immunology." *History of Science* 30 (1992): 371–95.

Lunbeck, Elizabeth. "Identity and the Real Self in Postwar American Psychiatry." *Harvard Review of Psychiatry* 8 (2000): 318–22.

Mackay, Ian R., "The 'Autoimmune Diseases' 40th Anniversary." *Autoimmunity Reviews* 1 (2002): 5–11.

———. "Autoimmunity: Paradigms of Burnet and Complexities of Today." *Immunology and Cell Biology* 70 (1992): 159–71.

———. "Autoimmunity Since the 1957 Clonal Selection Theory: A Little Acorn to a Large Oak." *Immunology and Cell Biology* 85 (2007): 1–5.

———. "Burnet and Autoimmunity." In *Walter and Eliza Hall Institute of Medical Research Annual Review: Special Volume, 1978–79, A Tribute to Sir Macfarlane Burnet,* vol. 38, Melbourne: Exchange Press, 1979, 39–45.

———. "The 'Burnet Era' of Immunology: Origins and Influence." *Immunology and Cell Biology* 69 (1991): 301–5.

———. "Historical Reflections on Autoimmune Hepatitis." *World Journal of Gastroenterology* 14 (2008): 3292–3300.

———. "Roots of and Routes to Autoimmunity." In *Immunology: The Making of a Modern Science,* edited by Richard B. Gallagher, Jean Gilder, G. J. V. Nossal, and Gaetano Salvatore, 49–63. London: Academic Press, 1995.

Mackay, Ian R., and Warwick Anderson. "What's in a Name? Experimental Encephalomyelitis: 'Allergic' or 'Autoimmune'? *Journal of Neuroimmunology* 223 (2010): 1–4.

Mackay, Ian R., and B. D. Tait. "The History of Autoimmune Hepatitis." In *Autoimmune Hepatitis,* edited by M. Nishioka, G. Toda, and M. Zeniya, 3–23. Amsterdam: Elsevier, 1994.

Maines, David. "The Social Arrangements of Diabetic Self-Help Groups." In *Chronic Illness and the Quality of Life.* 2nd ed., edited by Anselm Strauss, Juliet Corbin, Shzuko Fagenhaugh, Barney G. Glaser, David Maines, Barbara Suczek, Carolyn Wiener, 111–26. St. Louis: C. V. Mosby, 1984.

Mallavarapu, Ravi K., and Edwin W. Grimsley. "The History of Lupus Erythematosus." *Southern Medical Journal* 100 (2007): 896–98.

Marchalonis, John J. "Burnet and Nossal: The Impact on Immunology of the Walter and Eliza Hall Institute." *Quarterly Review of Biology* 69 (1994): 53–67.

Marks, Harry M. "Cortisone, 1949: A Year in the Political Life of a Drug." *Bulletin of the History of Medicine* 66 (1992): 419–39.

———. *The Progress of Experiment: Science and Therapeutic Reform in the United States, 1900–1990.* Cambridge: Cambridge University Press, 1997.

Markson, E. W. "Patient Semiology of a Chronic Disease: Rheumatoid Arthritis." *Social Science and Medicine* 5 (1971): 159–67.

Marquardt, Martha. *Paul Ehrlich.* London: William Heinemann, 1949.

Martin, Raymond, and John Barresi. *The Rise and Fall of Soul and Self: An Intellectual History of Personal Identity.* New York: Columbia University Press, 2006.

Maulitz, Russell C., and Diana E. Long, eds. *Grand Rounds: One Hundred Years of Internal Medicine.* Philadelphia: University of Pennsylvania Press, 1988.

Mazumdar, Pauline M. H. "The Purpose of Immunity: Landsteiner's Explanation of the Human Isoantibodies." *Journal of the History of Biology* 8 (1975): 115–34.

———. *Species and Specificity: An Interpretation of the History of Immunology.* Cambridge: Cambridge University Press, 1995.

Medvei, Victor C. *The History of Clinical Endocrinology: A Comprehensive Account of Endocrinology from Earliest Times to the Present Day.* New York: Informa Healthcare, 1993.

Mendelsohn, J. Andrew. "'Like all that Lives': Biology, Medicine and Bacteria in the Age of Pasteur and Koch." *History and Philosophy of the Life Sciences* 24 (2002): 3–36.

———. "Medicine and the Making of Bodily Inequality in Twentieth-Century Europe." In *Heredity and Infection: The History of Disease Transmission*, edited by Jean-Paul Gaudillière and Ilana Löwy, 21–79. London: Routledge, 2001.

Mitchell, W. J. T. "Picturing Terror: Derrida's Autoimmunity." *Critical Inquiry* 33 (2007): 277–90.

Mitman, Gregg. *Breathing Space: How Allergies Shape Our Lives and Landscapes.* New Haven: Yale University Press, 2007.

Moberg, Carol L., ed. *Entering an Unseen World: A Founding Laboratory and the Origins of Modern Cell Biology, 1910–1974.* New York: Rockefeller University Press, 2012.

Monks, Judith, and Ronald Frankenburg. "Being Ill and Being Me: Self, Body and Time in Multiple Sclerosis Narratives." In *Disability and Culture*, edited by Benedicte Ingstad and Susan Reynolds Whyte, 107–33. Berkeley: University of California Press, 1995.

Moulin, Anne Marie. "The Dilemma of Medical Causality and the Issue of Biological Individuality." In *Science, Technology and the Art of Medicine: European-American Dialogues*, edited by Corinna Delkeskamp-Hayes and Mary Ann Gardell Cutter, 153–62. Dordrecht: Kluwer, 1993.

———. "Fleck's Style." In *Cognition and Fact: Materials on Ludwik Fleck*, edited by Robert S. Cohen and Thomas Schelle, 407–19. Dordrecht: D. Reidel, 1986.

———. "The Immune System: A Key Concept in the History of Immunology." *History and Philosophy of the Life Sciences* 11 (1989): 221–36.

———. "La métaphore du soi et le tabou de l'autoimmunité." In *Soi et non-soi*, edited by J. Bernard, M. Bessis, and Claude Debru. Paris: Seuil, 1990.

———. *Le dernier langage de la médecine: Histoire de l'immunologie de Pasteur au Sida.* Paris: Presses Universitaires de France, 1991.

———. "Multiple Splendor: The One and Many Versions of the Immune System." In *Singular Selves: Historical Issues and Contemporary Debate in Immunology*, edited by Anne Marie Moulin and Alberto Cambrosio, 228–43. Amsterdam: Elsevier, 2001.

Mukherjee, Siddhartha. *The Emperor of All Maladies: A Biography of Cancer.* New York: Harper Collins, 2010.

Murray, T. Jock. "The History of Multiple Sclerosis: The Changing Frame of the Disease over the Centuries." *Journal of the Neurological Sciences* 277 (2009): 53–58.

———. *Multiple Sclerosis: The History of a Disease.* New York: Demos Health, 2005.

Naas, Michael. "'One Nation . . . Indivisible': Jacques Derrida on the Autoimmunity of Democracy and the Sovereignty of God." *Research in Phenomenology* 36 (2006): 15–44.

Nossal, Gustav J. V. *Diversity and Discovery: The Walter and Eliza Hall Institute 1965–1996.* Melbourne: Miegunyah Press, 2007.

Park, Hyung Wook. "Germs, Hosts, and the Origin of Frank Macfarlane Burnet's Concept of 'Self' and 'Tolerance,' 1936–1949." *Journal of the History of Medicine and Allied Sciences* 61 (2006): 492–534.

———. "The Shape of the Human Being as a Function of Time: Time, Transplantation, and Tolerance in Peter Brian Medawar's Research, 1937–1956." *Endeavour* 34 (2010): 112–21.

Parnes, Ohad. "'Trouble from Within': Allergy, Autoimmunity, and Pathology in the First Half of the Twentieth Century." *Studies in History and Philosophy of Biological and Biomedical Sciences* 34 (2003): 425–54.

Pauly, Philip P. *Controlling Life: Jacques Loeb and the Engineering Ideal in Biology.* New York: Oxford University Press, 1987.

Pelling, Margaret. *Cholera, Fever and English Medicine, 1825–1865.* Oxford: Oxford University Press 1978.

Peterson, Audrey C. "Brain Fever in Nineteenth-Century Literature: Fact and Fiction." *Victorian Studies* 19 (1976): 445–64.

Pickstone, John. *Ways of Knowing: A New History of Science, Technology and Medicine.* Chicago: University of Chicago Press, 2000.

Platts, Margaret M. "Some Medical Syndromes Encountered in Nineteenth-Century French Literature." *Medical Humanities* 27 (2001): 82–88.

Potter, Brian. "The History of the Disease called Lupus." *Journal of the History of Medicine and Allied Sciences* 48 (1993): 80–90.

Pradeu, Thomas. *The Limits of the Self: Immunology and Biological Identity.* New York: Oxford University Press, 2012.

Pradeu, Thomas, and Edwin L. Cooper. "The Danger Theory: 20 Years Later." *Frontiers in Immunology* 3 (2012): 1–9.

Quétel, Claude, *The History of Syphilis.* Translated by J. Braddock and B. Pike. Baltimore: Johns Hopkins University Press, 1992.

Rabinow, Paul. "Artificiality and Enlightenment: From Sociobiology to Biosociality." In *Essays on the Anthropology of Reason,* 91–111. Princeton: Princeton University Press, 1996.

Rasmussen, Nicolas. "Freund's Adjuvant and the Realization of Questions in Postwar Immunology." *Historical Studies in the Physical and Biological Sciences* 23 (1993): 337–66.

———. "Steroids in Arms: Science, Government, Industry, and the Hormones of the Adrenal Cortex in the United States, 1930–1950." *Medical History* 46 (2002): 299–324.

Rheinberger, Hans-Jörg. *On Historicizing Epistemology: An Essay*. Stanford: Stanford University Press, 2010.

———. *Toward a History of Epistemic Things: Synthesizing Proteins in the Test Tube*. Stanford: Stanford University Press, 1997.

———. "Translating Derrida." *CR: The New Centennial Review* 8 (2008): 175–87.

Robinson, I. "Personal Narratives, Social Careers and Medical Courses: Analyzing Life Trajectories in Autobiographies of People with Multiple Sclerosis." *Social Science and Medicine* 30 (1990): 1173–86.

Rose, Nikolas S. "How Should One Do the History of the Self?" In *Inventing Our Selves: Psychology, Power, and Personhood*. New York: Cambridge University Press, 1996.

Rose, Noel R., and Ian R. Mackay, eds. *The Autoimmune Diseases*, 5th ed. London: Elsevier, 2013.

Rosenberg, Charles E. "The Bitter Fruit: Heredity, Disease and Social Thought in Nineteenth-century America." *Perspectives in American History* 8 (1974): 189–235.

———. *The Cholera Years: The United States in 1832, 1849, and 1866*. Chicago: University of Chicago Press, 1962.

———. "Martin Arrowsmith: The Scientist as Hero." In *No Other Gods*, 123–32. Baltimore: Johns Hopkins University Press, 1976.

———. "The Therapeutic Revolution: Medicine, Meaning and Social Change in Nineteenth-Century America." *Perspectives in Biology and Medicine* 20 (1977): 485–506.

———. "The Tyranny of Diagnosis: Specific Entities and Individual Experience." *Milbank Quarterly* 80 (2002): 237–60.

Rothfield, Lawrence. *Vital Signs: Medical Realism in Nineteenth-Century Fiction*. Princeton: Princeton University Press, 1992.

Rowland, Lewis P. *The Legacy of Tracy J. Putnam and H. Houston Merritt: Modern Neurology in the United States*. Oxford: Oxford University Press, 2009.

Rubin, L. P. "Styles in Scientific Explanation: Paul Ehrlich and Svante Arrhenius on Immunochemistry." *Journal of the History of Medicine* 35 (1980): 397–45.

Sankaran, Neeraja. "The Bacteriophage, Its Role in Immunology: How Macfarlane Burnet's Phage Research Shaped His Scientific Style," *Studies in History and Philosophy of Biological and Biomedical Sciences* 41 (2010): 367–75.

———. "Frank Macfarlane Burnet and the Nature of Bacteriophage, 1924–1937." Ph.D. dissertation, Yale University, 2006.

Sapp, Jan. *Beyond the Gene: Cytoplasmic Inheritance and the Struggle for Authority in Genetics*. New York: Oxford University Press, 1987.

Schmalsteig, F. C., Jr., and A. S. Goldman. "Jules Bordet (1870–1961): A Bridge between Early and Modern Immunology." *Journal of Medical Biography* 17 (2009): 217–24.

Schorske, Carl E. *Fin-de-Siècle Vienna: Politics and Culture*, New York: Viking, 1981.

Sexton, Christopher. *The Seeds of Time: The Life of Sir Macfarlane Burnet.* Oxford: Oxford University Press, 1991.

Sigerist, Henry E. *Man and Medicine: An Introduction to Medical Knowledge.* London: George Allen and Unwin, 1932.

Silverstein, Arthur M. "The End Is Near! The Phenomenon of the Declaration of Closure in a Discipline." *History of Science* 37 (1999): 1–19.

———. *A History of Immunology.* San Diego: Academic Press, 1989.

———. *A History of Immunology*, 2nd ed. San Diego: Academic Press, 2009.

———. "Horror Autotoxicus, Autoimmunity, and Immunoregulation: The Early History." *Transfusion Medicine and Hemotherapy* 32 (2005): 296–302.

———. *Paul Ehrlich's Receptor Immunology: The Magnificent Obsession.* San Diego: Academic Press, 2001.

———. "Whatever Happened to Cell-Bound Antibodies? On the Overriding Influence of Dogma." *Nature Immunology* 3 (2002): 105–8.

Slater, Leo. "Industry and Academy: The Synthesis of Steroids." *Historical Studies in the Physical and Biological Sciences* 30 (2000): 443–80.

Söderqvist, Thomas. "Darwinian Overtones: Niels K. Jerne and the Origin of the Selection Theory of Antibody Formation." *Journal of the History of Biology* 27 (1994): 481–529.

———. *Science as Autobiography: The Troubled Life of Niels Jerne.* Translated by Daniel Mel Paul. New Haven: Yale University Press, 2003.

Soto Laveaga, Gabriela. *Jungle Laboratories: Mexican Peasants, National Projects, and the Making of Global Steroids.* Durham, NC: Duke University Press, 2009.

Stapleton, Darwin H., ed. *Creating a Tradition of Biomedical Research: Contributions to the History of the Rockefeller University.* New York: Rockefeller University Press, 2004.

Starobinski, Jean. *Action and Reaction: The Life and Adventures of a Couple.* Translated by Sophie Hawkes with Jeff Fort. New York: Zone Books, 2003.

Stenager, Egon. "The Course of Heinrich Heine's Illness: Diagnostic Considerations." *Journal of Medical Biography* 4 (1996): 28–32.

Stevenson, Lloyd G. "Exemplary Disease: The Typhoid Pattern." *Journal of the History of Medicine* 37 (1982): 159–81.

Stewart, D. C., and T. J. Sullivan. "Illness Behavior and the Sick Role in Chronic Disease: The Case of Multiple Sclerosis." *Social Science and Medicine* 16 (1982): 1397–1404.

Storey, G. O., M. Comer, and D. L. Scott. "Chronic Arthritis Before 1876: Early British Cases Suggesting Rheumatoid Arthritis." *Annals of the Rheumatic Diseases* 53 (1994): 557–60.

Strasser, Bruno J., and Bernardino Fantini. " Molecular Diseases and Diseased Molecules: Ontological and Epistemological Dimensions." *History and Philosophy of the Life Sciences* 20 (1998): 189–215.

Strauss, Anselm L., Juliet Corbin, Shzuko Fagenhaugh, Barney G. Glaser, David Maines, Barbara Suczek, and Caroline Wiener, eds. *Chronic Illness and the Quality of Life*, 2nd ed. St. Louis: C. V. Mosby, 1984.
Sturdy, Steve. "Looking for Trouble: Medical Science and Clinical Practice in the Historiography of Modern Medicine." *Social History of Medicine* 24 (2011): 739–57.
Summers, William C. "Hans Zinsser: A Tale of Two Cultures." *Yale Journal of Biology and Medicine* 72 (1999): 341–47.
Swann, John Patrick. *Academic Scientists and the Pharmaceutical Industry: Cooperative Research in Twentieth-Century America.* Baltimore: Johns Hopkins University Press, 1988.
Talley, Colin L. "The Emergence of Multiple Sclerosis as a Nosological Category in France, 1838–1868." *Journal of the History of the Neurosciences* 12 (2003): 250–65.
———. "The Emergence of Multiple Sclerosis, 1870–1950: A Puzzle in Historical Epidemiology." *Perspectives in Biology and Medicine* 48 (2005): 383–95.
———. *A History of Multiple Sclerosis.* New York: Praeger, 2008.
———. "The Treatment of Multiple Sclerosis in Los Angeles and the United States, 1947–1960." *Bulletin of the History of Medicine* 77 (2003): 874–99.
Tansey, E. M., S. V. Willihof, and D. A. Christie, eds. *Self and Non-Self: A History of Autoimmunity.* London: Wellcome Institute for the History of Medicine, 1997.
Tauber, Alfred I. *The Immune Self: Theory or Metaphor?* Cambridge: Cambridge University Press, 1994.
———. "The Immunological Self: A Centenary Perspective." *Perspectives in Biology and Medicine* 35 (1991): 74–86.
———. "Immunology and the Enigma of Selfhood." In *Growing Explanations: Historical Perspectives on Recent Science*, edited by M. Norton Wise, 201–21. Durham, NC: Duke University Press, 2004.
———. "The Molecularization of Immunology." In *The Philosophy and History of Molecular Biology: New Perspectives,* edited by Sahotra Sarkar, 125–69. Dordrecht: Kluwer, 1996.
Tauber, Alfred I., and Leon Chernyak. *From Metaphor to Theory: Metchnikoff and the Origin of Immunology.* New York: Oxford University Press, 1991.
Tauber, Alfred I., and Scott H. Podolsky. "Frank Macfarlane Burnet and the Immune Self." *Journal of the History of Biology* 27 (1994): 531–73.
———. *The Generation of Diversity: Clonal Selection Theory and the Rise of Molecular Immunology.* Cambridge, MA: Harvard University Press, 1997.
Taylor, Charles. *Sources of the Self: The Making of the Modern Identity.* Cambridge, MA: Harvard University Press, 1989.
Temkin, Owsei. "The Scientific Approach to Disease: Specific Entity and Individual Sickness." In *The Double Face of Janus and Other Essays in the History of Medicine,* Baltimore: Johns Hopkins University Press, 1977.

———. "Wunderlich, Schelling and the History of Medicine." In *The Double Face of Janus and Other Essays in the History of Medicine.* Baltimore: Johns Hopkins University Press, 1977.

Tracy, Sarah W. "An Evolving Science of Man: The Transformation and Demise of American Constitutional Medicine, 1900–1950." In *Greater Than the Parts: Holism in Biomedicine, 1920–1950,* edited by Christopher Lawrence and George Weisz, 161–88. Oxford: Oxford University Press, 1998.

Vidal, Fernando. "Brains, Bodies, Selves, and Science: Anthropologies of Identity and the Resurrection of the Body." *Critical Inquiry* 28 (2002): 930–74.

Viner, Russell. "Putting Stress in Life: Hans Selye and the Making of Stress Theory." *Social Studies of Science* 29 (1999): 391–410.

Wallace, Daniel J., and Ilana Lyon. "Pierre Cazenave and the First Detailed Modern Description of Lupus Erythematosus." *Seminars in Arthritis and Rheumatism* 28 (1999): 305–12.

Waller, John C. "'The Illusion of an Explanation': The Concept of Hereditary Disease, 1770–1870." *Journal of the History of Medicine and Allied Sciences* 57 (2002): 410–48.

Ward, Candace. *Desire and Disorder: Fevers, Fictions, and Feeling in English Georgian Culture.* Lewisburg, PA: Bucknell University Press, 2007.

Weindling, Paul. "Scientific Elites and Laboratory Organization in Fin-de-Siècle Paris and Berlin: The Pasteur Institute and Robert Koch's Institute for Infectious Diseases Compared." In *The Laboratory Revolution in Medicine,* edited by Andrew Cunningham and Perry Williams, 170–88. Cambridge: Cambridge University Press, 1992.

Wiener, Carolyn L. "The Burden of Rheumatoid Arthritis: Tolerating the Uncertainty." *Social Science and Medicine* 9 (1975): 97–104.

Wilson, David. *The Science of Self: A Report on the New Immunology.* London: Longman, 1971.

Wilson, Leonard G. "Fevers and Science in Early Nineteenth Century Medicine." *Journal of the History of Medicine* 33 (1978): 386–407.

Winslow, Charles-Edward A. *The Conquest of Epidemic Disease: A Chapter in the History of Ideas.* Princeton: Princeton University Press, 1943.

Wittgenstein, Ludwig. *Philosophical Investigations,* 2nd ed. Translated by G. E. M. Anscombe. Oxford: Blackwell, 1958.

———. "Philosophy [from The Big Typescript, 1936]." In *Ludwig Wittgenstein: Philosophical Occasions, 1912–1951,* edited by J. Klagge and A. Nordman, 158–99. Indianapolis: Hackett, 1993.

Wood, Jane. *Passion and Pathology in Victorian Fiction.* Oxford: Oxford University Press, 2001.

Worboys, Michael. *Spreading Germs: Disease Theories and Medical Practice in Britain, 1865–1900.* Cambridge: Cambridge University Press, 2000.

INDEX